Astronomer's Pocket Field Guide

For other titles published in this series, go to
www.springer.com/series/7814

Charles A. Cardona III

Star Clusters

A Pocket Field Guide

with astronomical images by Jan Wisniewski

Charles A. Cardona III
Calverton, NY,
USA
cc@dreamcatchertechnology.com

ISBN 978-1-4419-7039-8 e-ISBN 978-1-4419-7040-4
DOI 10.1007/978-1-4419-7040-4
Springer New York Dordrecht Heidelberg London

Printed on acid-free paper

Springer is part of Springer Science+Business Media (www.springer.com)

This book is dedicated to my wife Kristin
and my boys; Christopher, Nicholas and Matthew,
who always inspire me in the way the mysteries
of the universe do

About the Author

Charles Cardona has been a teaching assistant in Astronomy at SUNY Suffolk Community College in New York, a former observatory director, a Variable Star section director, and a chairman of the board at the Custer Institute Observatory. He also taught courses, lectures, and seminars in Astronomy, Optics, and Computer Science. He was also the publisher of the Observatory Report Newsletter.

In addition to his astronomy work, Charles has been an entrepreneur and has built and sold several companies. During the past 10 years, he has donated to astronomy, education, and related children's causes. He recently organized the installation of a new observatory dome and telescope equipment.

He is currently involved in various educational programs and groups dedicated to finding missing children.

Preface

I have spent many wonderful nights observing at a variety of places around the world, but many of my fondest memories come from those made at the Custer Observatory in Southold on Long Island, where I grew up. I was only perhaps a dozen years old when I started out, and now more than three decades later I have logged *millions* of miles on the various telescopes there.

Some of the best memories are of observing and discussing astronomical topics with the many really terrific people who make up Custer. It truly holds a special place in my heart.

Star clusters were always among my favorite objects to observe and discuss there. This book is the result of the exposure to these objects, many of which I observed for the first time at Custer.

The Custer Institute Observatory was founded in the 1920s by local resident Charles W. Elmer of Perkin-Elmer Corp. fame. The original crew met at his house in Cedar Beach a few miles down the road from the institute. In the 1930s the group built an observatory complete with Mr. Elmer's donated 5" Alvan Clark refractor. Later the observatory expanded to include a 6" Clark refractor and other even larger instruments.

Among the notables who have been members and contributors were other local residents, famous optical engineer Laurits "Dan" Eichner, telescope maker Harry Fitz, physicist Albert Einstein, renowned astronomer Peter Van De Kamp, famous astronomer Sara Lee Lippincott, astronomer and author Fred Hess, Allen Seltzer of the Hayden Planetarium, instrument maker Bob Deroski, and many, many more.

Custer is a rustic brick building with a library, meeting hall, observatory dome, optical shop, and a quaint meeting room called the Elmer room. The Elmer Room is a paneled room complete with wood bookshelves surrounding a big brick fireplace. Many an intense conversation has been had in this room on cloudy nights and on cold clear ones while warming up. I have often sat up with several other members, way in to the wee hours of the night, discussing various astronomical topics, sometimes with the obligatory calculations on the blackboard to emphasize the point. Most of these wonderful discussions were accompanied by interesting books that covered some relevant topic or other. One such topic was star clusters, and I recall the lack of a book on this topic. This perhaps is what has prompted me to always wish to write on this topic.

This book is a guide to star clusters for amateur observers. It contains information on the brighter star clusters that are within reach of amateur telescopes from 3” to 30” in the northern latitudes. It was designed to guide amateurs through the interesting world of star clusters. It contains a concordance on their history, structure, features, observing, and photography.

Star clusters are diverse and interesting. Each has its own unique personality. And yet for some reason, as a topic of their own, they have been neglected. This book is an attempt to remedy that. I have attempted to provide information, but also a few interesting (to me, anyway) anecdotes and personal observations.

I certainly hope you enjoy the book. I have thoroughly enjoyed compiling it. Perhaps there will be a few animated discussions on star clusters prompted by it at Custer and other wonderful places as well.

Charles A. Cardona III
Calverton, NY

Acknowledgements

I would like to especially thank Jan Wisniewski, an amateur astrophotographer, who took most of the astronomical images that appear in this book. He has created truly beautiful images with equipment available to many amateurs. His tireless braving the cold nights is an inspiration to amateur and professional astronomers alike.

Contents

Dedication v

About the Author vii

Preface ix

Acknowledgements xi

Part 1 Background

1. Introduction 3
2. Understanding Stars and Star Clusters 5
3. How to Use This Book 17

Part 2 Star Cluster

4. Descriptions, Images, and Charts 21

Index 171

PART 1

BACKGROUND

CHAPTER 1

INTRODUCTION

Star clusters are collections of stars that have a shared gravitational bond. They range from loose aggregations of a few stars to mighty spherical-shaped globular clusters containing a million or more stars. Obviously the more stars and the more tightly packed the greater the gravitational attraction between them.

Studying star clusters is an important endeavor because it gives a look at stars of the same general age. This allows astronomers to see various elements of stellar evolution in progress from a common origin.

Some clusters have been known since ancient times and have special names. Many of these names have a mythological context, such as the Pleiades and the Hyades. Most of the brightest clusters are listed in Charles Messier's famous list of "comet-like" objects compiled in the 1700s. For some reason a few were not included in his list but are given a designation in the New General Catalog (NGC) or its supplement, the Index Catalog (IC) compiled by Danish astronomer J. L. E. Dreyer in the late 1800s and early 1900s.

Star clusters are some of the most interesting and beautiful objects that can be seen by the observer. Regardless of telescope size, from 3" to 30", there are many star clusters available that will provide stunning viewing. Some particularly beautiful star clusters are visible all year round and are ideal for both visual and photographic astronomical study.

Star clusters come in different varieties, densities, and sizes, each one having its own unique personality. From mighty globular clusters such as M-13 to the sweeping beauty of open clusters such as the Pleiades, each cluster has its own story to tell. Some are as old as the galaxy, some are new, and some are even being born right now.

C.A. Cardona III, *Star Clusters: A Pocket Field Guide,* Astronomer's Pocket Field Guide, DOI 10.1007/978-1-4419-7040-4 _1,

Many of the star clusters cataloged herein still have their formation nebula visible. This is the envelope of gas and dust from which the stars in the clusters formed. The youngest, of course, have the most gas and dust still around. In many cases these clusters are still in the process of star formation.

CHAPTER 2

UNDERSTANDING STARS AND STAR CLUSTERS

Stars are formed from gas and dust compressed together by various forces. Star clusters are formed in several different ways. In what are called open clusters, it is usually the sweeping wave action of the spiral arms of the galaxy thrusting vast stretches of this interstellar material away from the galaxy. In the case of what are known as globulars, it is believed that they formed during the original gravitational collapse that formed the galaxy. This is why the globulars are known to be much older than the open clusters.

Open clusters and globular clusters are the two main types of clusters. Each is related by virtue of the fact that they are collections of anywhere from a dozen to a million or more stars. The two types differ in structure, age, and distribution in the galaxy.

Stars are huge nuclear furnaces, converting their supplies of hydrogen to helium and eventually to heavier elements. This process takes place over the millions and billions of years of the stars' life cycle.

Early on, it was believed that stars were huge balls of burning gas. This theory could certainly account for the light and heat of a star, but it could not account for much more than a few thousand years' of a star's lifespan. From looking at the Sun and at fossils on Earth, we know that the Sun must have been burning for many millions of years. Several other theories were put forth, none of which could explain this discrepancy. Finally, with the discovery of radioactivity, a process was found that could provide the necessary energy to power a star for the millions or even billions of years required.

Stars are formed as vast clouds of gas and dust slowly contract under their own gravitational attraction. This process accelerates as it progresses, due to the increasing gravitational attraction of the increasing mass. If the mass is large enough the temperature in the center rises until enough compression energy causes hydrogen atoms to be squeezed together and form helium atoms. The amount of energy released from just 1 g of hydrogen

C.A. Cardona III, *Star Clusters: A Pocket Field Guide,* Astronomer's Pocket Field Guide, DOI 10.1007/978-1-4419-7040-4_2, being converted to helium is 6.4×10^{18} ergs of energy. In the Sun, a rather modest star, 4×10^{33} ergs of energy are being produced each second. This is equivalent to the energy produced by 6 trillion Hiroshima-sized bombs exploding each second. It means that each second 600 million tons of hydrogen is converted to helium, and more than 4 million tons of matter is converted to energy each second.

Although this seems an extremely high rate of consumption, if the Sun were to convert even half its mass of 2×10^{33} g at that rate, it would take more than 10 billion years to use it up. This clearly is the source of the Sun's and other stars' energy.

The rate at which stars burn energy is related to their mass. The larger the star the higher its internal compression, and the faster it burns its fuel. In fact, the rate of burning increases far in excess of the increase in mass; therefore large stars use up their fuel much quicker than small ones.

As the star compresses from its original cloud it finally begins to shine forth, creating a solar wind, which eventually disperses the cloud from which it formed. This process occurs throughout the cloud of gas and dust, and on most occasions more than one object is formed. Some will be small and merely planets, while others will be larger and become stars. Such groups of stars formed together usually stay together, at least for a while. These are star clusters.

Astronomers are always looking for ways in which to study and classify astronomical objects. One of the best ways to categorize stars is by their spectra. The stars are categorized in various spectral types, or colors. This classification system was developed at Harvard Observatory to help astronomers classify and understand the different types of stars. The types are O, B, A, F, G, K, and M. There are also a few others, which are sub-classes of K. The O stars are the hottest and whitest and the M stars are the coolest and reddest. Each class is further broken down into ten subdivisions 0 through 9. Our Sun is a G2 star, a rather average semi-cool yellow star.

The spectra of stars are not actually continuous, like a rainbow, but with careful examination have distinct bright lines where specific colors are emitted and areas in between where those colors are not emitted at all. Each star has its own unique spectrum, or signature.

Electron orbital motions cause the spectral lines. Electrons are small particles with a negative charge that orbit the atomic nucleus. Electrons must always move in distinct orbit levels. If the electron moves to the next lower orbit it loses energy and spits out a light particle (photon) of a specific color (energy). Each different atom has different color photons, which are emitted through various processes. Through experimentation in the laboratory, scientists have cataloged the spectra of all the various elements and therefore can use this as a guide to knowing the composition of the furthest stars.

The benefits of stellar spectra don't stop there. Since light has a wave structure it is subject to the Doppler effect. This effect is familiar to anyone who has heard a train whistle rise and then fall as the train passes by. The sound waves are compressed as they approach and stretch apart as they recede. This effect, sometimes known as red shift or blue shift, is also present in stellar spectra if the object is moving either towards or away from us. The spectra of receding objects shifts toward the red end, and that of an approaching object shifts toward the blue end. By careful measurement of this deviation astronomers can determine quite accurately the recession or approach of a star (its radial velocity). Positive radial velocity means the star is approaching us, and negative radial velocity means the star is moving away from us.

This effect also is sometimes seen to vary, as if the star sometimes approaches and sometimes recedes relative to its normal motion. Astronomers know that this object is orbiting another object or objects, and its motion helps astronomers to identify double- and multiple-star systems. Many of these stars are much too close together to see as separate objects even in the biggest telescopes. These objects are known as spectroscopic binaries. Many thousands of such multiple systems have been identified since this technique was first used.

The spectra of stars are also useful in determining the age of stars. The most useful tool in determining star evolution is the Hertzsprung–Russell (HR) diagram, which is a graph showing temperature versus magnitude (brightness). The HR diagram was developed in the early twentieth century by Ejnar Hertzsprung and Henry Norris Russell as a method to classify stellar evolution.

If we plot all the stars on this graph, we can see that most of them fall along a diagonal line from the upper left to the lower right. This is known as the

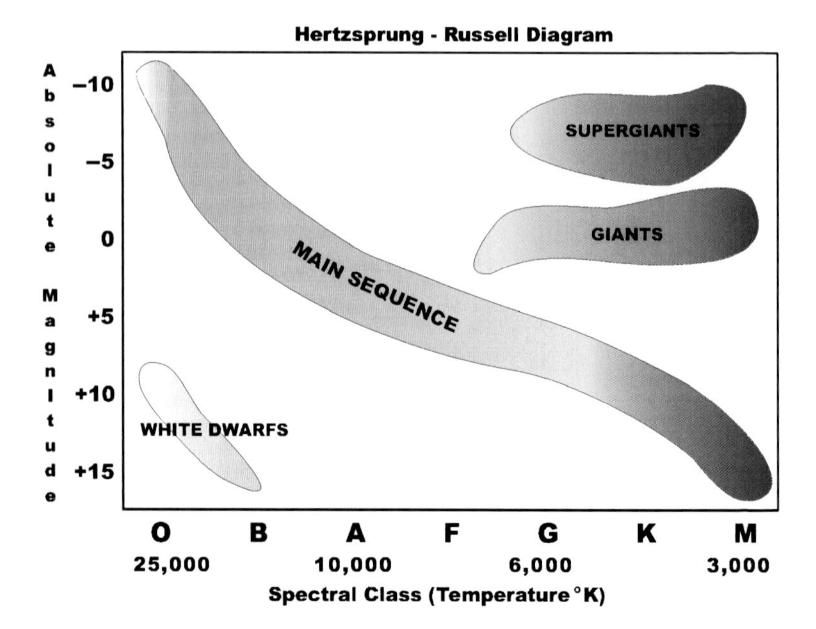

main sequence. It is believed that stars begin their life at the upper right portion of the graph and then move to their place on the main sequence shortly after they begin to burn their nuclear fuel. Their place on the main sequence is dictated by their mass. The higher the mass the further to the left on the main sequence they fall.

A star spends much of its life in the main sequence part of the diagram. Our Sun has been on the main sequence for 5 billion years and will continue there for another 5 billion. The outward pressure of the energy produced in the core is balancing the inward pressure of gravity. This "balancing act" is what allows the star to stay relatively stable on the main sequence for all that time.

As a star begins to burn up all its hydrogen fuel, the energy produced is reduced, allowing the core to contract; this contraction causes more heat to be generated, which ultimately puffs up the outer layers of the star. The star begins to move off the main sequence and, depending on how much mass it has, will determine its ultimate fate. Smaller stars such as the Sun will go through various death throes as they slowly run out of hydrogen. As the core contracts, the temperature will rise enough for helium to begin burning in the core. This higher temperature will push the star's outer layers further away, creating a red giant stage. The star will usually go through several violent episodes during this stage and produce large nebulae from explosions and stellar wind. The star will then slowly settle down and fade away as a white dwarf.

Larger stars will explode as novae and produce beautiful planetary nebulae such as M27 and M57. The largest stars may even go supernova and explode, destroying themselves leaving nothing but a neutron star or black hole as a reminder of their former glory.

We can tell the age of a star cluster by looking at the types of stars that make it up. If a cluster contains many white giant stars, we know it cannot be very old. In fact, by plotting an HR diagram of the stars in a particular cluster, we can identify the youngest stars that make it up. This is one way in which astronomers can help to determine the age of our galaxy.

All these tools are very useful for astronomers who wish to study star clusters. Stellar spectra help to provide the age and composition of the star cluster. It also can help in inferring the original composition of the formation nebulae.

Open Star Clusters

Open star clusters have been observed as such for thousands of years, and some are part of astronomical lore in various cultures.

Open star clusters are found primarily in the galactic disk and are much younger than globular clusters. The stars in open clusters are typical Population I stars; this means they are young and rich in metals. The Population I stars are typically located in the disk of the galaxy and were formed recently (astronomically speaking) from the gas and dust swept around by the spiral arms. The gas and dust in these clouds are enriched with metals and other heavy elements from the explosions of countless supernovae.

As this gas is swept around in the spiral arms of the galaxy, it is compressed. This compression action on the gas and dust causes pools and eddies to form, which are known as nebulae, and among these swirling collections of gas, stars begin to form. Some nebulae can condense and create dozens, some even thousands, of stars. These stars are still bound together gravitationally and move together in an elaborate dance as they circle the galaxy. These groups of stars are what we see as open clusters. They typically have little defined form and can contain from a few to thousands of stars. Some, such as the Pleiades (M45) and M16, actually have some of the nebula they were formed from still surrounding them.

Open star clusters are classified using a methodology devised by Robert Julius Trumpler, a renowned Swiss astronomer of the early twentieth century. The classification is divided into three parts: density, luminosity function, and number of stars. The density is denoted by Roman numerals from I to IV, with I being the most concentrated clusters and IV being the loosest, with almost no structure or concentration. The luminosity function, which is a numeral, has to do with the number of bright stars in the clusters and their luminosity. The final element is the letter p, m, or r to denote the total number of stars in the cluster. The letter p denotes poor – clusters with less than 50 stars. The letter m denotes medium rich clusters with 50–100 stars, and r denotes rich clusters with over 100 stars. For example the Pleiades (M45) is classified as I 3 r, meaning well concentrated with fairly luminous stars and over 100 members.

Open clusters are not all stable; in fact many will drift apart over several millions of years as tidal forces from different sections of the galaxy slowly pull the stars apart. The Sun probably formed in a cluster similar to the Pleiades, and over many millions of years the various members have been strewn throughout the spiral arms in our galaxy. There are calculations which show that clusters with less than one star per cubic parsec become unstable very quickly. The Hyades is an example of a cluster with a low density (1 star per 40 cubic parsecs) and is therefore very unstable. With a density of about 1 star per 10 cubic parsecs the Pleiades is much more stable and therefore should take much longer to dissipate. With this knowledge we know that clusters are relatively young objects in our galaxy.

Open clusters are categorized in many well-known catalogs. In the 1700s Charles Messier began his famous "Messier list" of fuzzy objects, so observers wouldn't be confused when searching for new comets. Many noted astronomers of his day were somewhat superstitious. As all intelligent people of the time knew that comets portended grave events, it would only make sense to have effective methods to search for such objects. Thus Charles Messier compiled his list of comet-like objects, which included a great number of star clusters. As he and others of his era looked at these objects carefully in their telescopes they discovered that many were in actuality dozens or even hundreds of individual stars. These groups of stars were compressed together in such a fashion that they could be nothing but far off families of stars traveling through the galaxy together.

Several others compiled lists including star clusters, notably Nicolas Louis de Lacaille, Edmund Halley, William Herschel, and Harlow Shapley. More modern catalogs include the IC, NGC, Atlas Coeli, and probably foremost *The Catalog of Star Clusters*, which provides data and details on over 1,000 open clusters.

Open star clusters are often tenuous and filamentary objects, which vary greatly in structure. Some are tight knots of stars, others loose associations. Each has its own unique personality, providing the observer with many nights of enjoyment. Observing deep into the central regions of star clusters can be almost an entrancing experience. Imagine what it must be like to be on a planet circling one of the stars in that cluster, to look up and see a sky full of brilliant blue young stars, and red and yellow giant stars blazing high in the sky! Perhaps some wisps of the nebula that formed the more recent stars still show. Undoubtedly a wondrous experience.

Globular Star Clusters

Globular star clusters were certainly seen by ancient observers. However, until the advent of the telescope it wasn't realized that they were actually clumps of hundreds or even thousands of stars. Early observers called them nebulae.

Among the first astronomers to record observations of globular star clusters were German astronomer Johannes Hevelius in the seventeenth century and Edmund Halley. They and others discovered that these objects were not simply nebulae but were celestial cities of stars. Thousands and thousands of stars condensed into tight little "globes" (hence the word "globular") of stars. As time passed, many noted astronomers such as William Herschel, Charles Messier, and Abbe Lacaille discovered new globular clusters, until now we know of more than a hundred in our galaxy.

It is now known that globular star clusters are not limited to the galactic disk but are evenly distributed in a huge sphere around the galaxy known as the galactic halo. This halo is nearly twice the diameter of the galactic disk. The stars in globular star clusters were formed from the original gas and dust that formed our galaxy billions of years ago. These stars are known as Population II stars and are very poor in metals, as they were formed mostly from gas, which did not have the benefit of enrichment by supernova explosions.

Since globular star clusters are very old we can expect the stars making them up to be old also. The best estimates show them to be 12 or more billion years old. In fact, cosmologists who have been trying to accurately determine the age of the universe have closely studied globular clusters. Since these clusters have been circling the galaxy for billions of years, they have undoubtedly been subject to a variety of gravitational disturbances, which has upset the internal balance and probably has caused numerous stars to be ejected. This may account for the many lone Population II stars distributed in the galactic halo.

The stars inside globular star clusters perform a complicated dance together as they orbit the center of gravity of these immense objects. However, numerous encounters and collisions occur, especially in the densest clusters. The stars, although bound together by the intense gravitation of the cluster, follow extremely complex orbits, spending some of their time at the outer fringes and some near the core.

Globular cluster halos similar to the one in our galaxy have been discovered in other galaxies. The Tarantula nebula in the Large Magellanic Cloud (a satellite galaxy of our own) is believed to be the birthing place of a future globular cluster.

Globular clusters commonly contain variable stars of the RR Lyrae type (also known as cluster-type variables). These are large stars, which are moving off the main sequence and have begun to pulsate. They are roughly 6 times the mass and 50 times the luminosity of the Sun. They are characterized by pulsation periods of less than a day and variation of about one magnitude. They are somewhat similar to the Cepheid variables in that they pulsate regularly. As with the Cepheids there is a relationship between the pulsation period of RR Lyrae stars and their luminosity, and so they are useful as distance indicators. This is how the distances to many of the globular clusters have been more accurately refined. Unfortunately not all globular star clusters contain these variables.

There are a large number of globular star clusters that have been found to have pulsars in their cores. In many cases more than a dozen pulsars have been found in a single globular cluster. Pulsars are rapidly rotating neutron stars that are the result of a supernova explosion. In the final years of a star several times the mass of the Sun, the star runs out of hydrogen in its core. The core loses its outward pressure from the fusion of hydrogen and begins to compress. This compression causes the core to heat up. Eventually the temperature gets high enough to begin helium fusion. The outer layers surrounding the core still contain hydrogen and are hot enough for hydrogen fusion to occur. This larger "hot" area causes the outer layers of the star to swell up enormously. Since the surface area increases by the square of the diameter, the outer layers become much cooler than before. In cooling off, they become redder. The star has now become a red giant.

The helium in the core fuses into carbon, oxygen, and nitrogen. This process produces much less energy than hydrogen fusion and thus depletes the helium much quicker. The various atomic nuclei fuse as the core desperately condenses and drains itself out of energy faster and faster. Finally iron is produced. Iron cannot undergo the type of fusion that releases energy. In fact fusion of iron consumes energy. Iron becomes a nuclear dead end for the star.

At this point the star keeps collapsing, with no energy left to restrain it. This collapse happens very rapidly, in a matter of hours. As this collapsing shell implodes upon itself, the large quantities of hydrogen in the surrounding shell slam down and fuse suddenly in a matter of minutes, causing a huge explosion that blows the outer layers of the star outward.

After the explosion, the core continues to collapse and becomes a soup of compressed protons, neutrons, and electrons. The repulsion of the negative electrons and the positive protons is so great that the star's compression is stopped at about the size of a planet. This compressed star usually becomes a white dwarf. The white dwarf will continue to slowly cool over many billions of years until it finally becomes a burnt-out cinder known as a black dwarf.

There are other possibilities, though. In the 1930s Indian-American astronomer S. Chandrasekhar calculated that if a star was more massive than 1.44 times the mass of the Sun the repulsion of the protons and neutrons wouldn't be enough to stop the collapse. This number is known as "Chandrasekhar's limit." The star squeezes the electrons and protons together to form neutrons, and the star becomes nothing but a soup of neutrons. This highly compressed "neutron star," as it is known, is less than 20-miles in diameter.

One interesting feature of neutron stars is that they rotate very rapidly, in many cases multiple times every second and in some hundreds of times a second. The reason for this comes from the law of conservation of angular momentum. The original star was rotating, perhaps once a month or so. However, the star was much, much larger, as the star contracted, angular momentum had to be conserved, and so the star spun faster. This is demonstrated when figure skaters draw their arms in and spin faster.

As these stars spin very rapidly, they generate huge magnetic fields that spew radiation from the magnetic poles. These poles apparently are not always located at the axes of rotation. This produces a flashing beacon, which if pointed at or near us can be seen as a flashing radio signal. They have also been detected flashing in visual light.

Pulsars are extremely accurate time keepers and pulsate more regularly than even the finest atomic clocks here on Earth. The rotations of pulsars do slow down ever so slowly over millions of years, however, as the particles streaming away carry tiny bits of the angular momentum away with them.

Even larger stars theoretically will compress so far that even the neutrons collapse, and the star becomes a soup of quarks. The density of such an object is so high that the acceleration of gravity close to it is above the speed of light. This object is known as a black hole. Black holes can only be detected indirectly. If they are located near other stars or other sources of gas and dust, they will suck them in like a cosmic vacuum cleaner.

As the gas gets sucked inside the black hole, it is accelerated so fast that it emits high-energy radiation in the form of X-rays and gamma rays.

It is deep within the cores of globular clusters that astronomers are now looking for pulsars and black holes. This field of study is revealing many interesting new details, including the recent discovery of gas clouds inside globular clusters.

Telescopes for Observing Star Clusters

There are many fine star clusters, both open and globular, available to a wide variety of instruments, from binoculars to large aperture reflectors. A few nice open clusters look quite striking in binoculars. However the magnification and light-gathering ability of even a 3″ telescope can bring many dozens of beautiful clusters into view.

A good number of nice clusters are brighter than ninth magnitude, which makes them easily visible even with the small 'scope.

Many clusters are nicely within reach of the amateur astrophotographer, especially with the new, highly accurate computer-controlled drive-correcting mounts available. With fast film and telescopes of F8 and below, nice results are possible on many objects with exposures of even 10–20 min. There are also many great possibilities with CCD cameras and image enhancement software. This new technology allows the astrophotographer to attain results previously only available to professionals with expensive equipment. In recent years CCD photography has almost all but taken over from film photography as the standard. One of the reasons for this is the more "linear" response of CCD technology as compared to film. This means that measurements of an object's brightness is much more accurate with CCD than with film.

CHAPTER 3

HOW TO USE THIS BOOK

This book is arranged by season in Right Ascension order from spring (from RA = 8 h) to winter (up to RA = 7 h). Of course the seasons overlap, so there will always be plenty to observe.

Included are some of the brightest and best clusters for amateur-sized telescopes and binoculars. Anyone with a telescope of 3″ (80 mm) or larger should be able to observe them all. Of course some will look a bit better in larger 'scopes than others, and some are best at low power. For imaging it is suggested to use a telescope of at least 5″ aperture to get the best results. Many of the images shown in this book have been taken with a standard 8″ Schmidt–Cassegrain telescope and amateur equipment.

Not every possible cluster is included. Some people will rue the omissions. However, each of those included has some special place in the author's observing heart, as I hope they will for you.

C.A. Cardona III, *Star Clusters: A Pocket Field Guide,* Astronomer's Pocket Field Guide, DOI 10.1007/978-1-4419-7040-4_3,

PART 2

STAR CLUSTER

CHAPTER 4

DESCRIPTIONS, IMAGES, AND CHARTS

M103 in Cassiopeia

Type: Open cluster
Designation(s): Messier 103, NGC 581
RA: 01h 33m
Dec: +60°39′f
Visual magnitude: 7.1
Distance: 8,100 LY
Size: 11′

M103 is a beautiful open cluster located in Cassiopeia. It is just beyond visual view but is wonderful in binoculars and seems to have a pronounced sense of depth. There is considerable contrast both in magnitude and color in the stars in M103, which somewhat small size makes it ideal for 50–100

C.A. Cardona III, *Star Clusters: A Pocket Field Guide,* Astronomer's Pocket Field Guide, DOI 10.1007/978-1-4419-7040-4_4, © Springer Science+Business Media, LLC 2010

power in a 3″ or bigger telescope. In some ways it resembles a wedge of cheese. Its triangular shape is one of its distinguishing characteristics. The Sun probably formed in a cluster similar to M103, which has now all but completely dispersed, as M103 surely will disappear in the next several billion years.

This cluster is quite young at only 20–25 million years. Many of the largest stars are still present, with one very noticeable red giant at tenth magnitude. The cluster is located more than 8,000 light years away and is at least 15 light years in diameter. It has a radial velocity of about 35 km/s in approach. Although not the richest cluster in the heavens, it contains more than 40 members and is very beautiful to behold.

Photographically M103 is a subject for eyepiece projection. Its smallish size makes it a good target with some magnification. There are however three other smaller clusters less than 2° away, which in a wider field picture are quite nice.

M103 is located about 1° north and east of the second magnitude blue star δ (delta) Cassiopeiae (Ruchbah). There are three other clusters - NGC 654, 659, and 663 about 2° to east towards ε (Epsilon) Cassiopeiae (Segin).

Suggested Instruments

binoculars
2″+ refractor
3″+ reflector
3″+ plus catadioptric

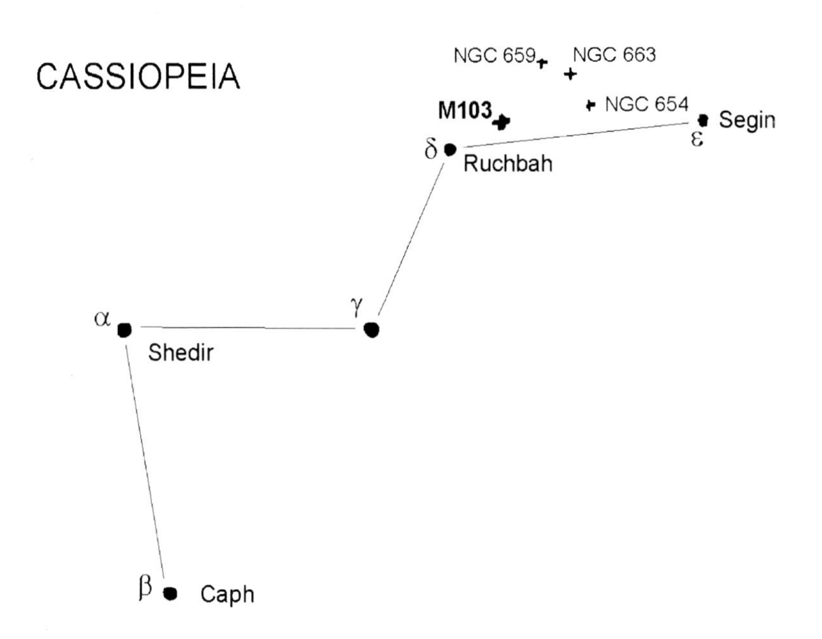
CASSIOPEIA
NGC 659
NGC 663
M103
NGC 654
Segin
ε
δ
Ruchbah
γ
α
Shedir
β
Caph

NGC 752 in Andromeda

Type: Open cluster
Designation(s): NGC 752
RA: 01h 58m
Dec: +37°41′
Visual magnitude: 6.2
Distance: 1,300 LY
Size: 60′

NGC 752 is a beautiful open cluster located just below γ (Gamma) Andromedae. It contains more than 100 stars with more than a dozen over tenth magnitude. This is a terrific object in binoculars.

Originally cataloged by William Herschel on September 21, 1786, NGC is believed to have been discovered by Giovanni Batista Hodierna, an astronomer in the court of the Duke of Montechiaro around 1650. He was one of the first to create a catalog of celestial objects with a telescope. He used a crude Galilean refractor of roughly 20 power. His book *De Admirandis Coeli Characteribus* contained a list of approximately 40 deep sky objects.

Widely scattered and old, this cluster contains a few dozen orange giant stars plus many main sequence stars. Perhaps 1.5 billion years in age, NGC-752 is in the process of drifting apart. Many of its members are barely holding on at the edge of the gravitational field. The members of this cluster are mostly G and K white stars with some larger orange and blue giants. It is believed this cluster formed at the outer edge of the galaxy where the dust and gas is thinner and lower in metal concentrations. NGC-752 is approaching us at 2.5 miles/s and is located at least 1,300 light years distant.

Photographically NGC-752 is an interesting challenge. It can be imaged using wider field CCD imagery with very satisfying results. The colorful aspect of this cluster due to its brilliant orange giant members makes it ideal for color photography.

Just 5° south of γ (Gamma) Andromedae (Almach), NGC-752 is quite easy to find. It is almost midway between Almach and α alpha Trianguli (Ras al Muthallath). It can be located as a fuzzy collection almost resolvable into stars in binoculars – resolving out nicely in a small telescope.

Suggested Instruments

binoculars
2.5″+ refractor
4″+ reflector
3″+ catadioptric

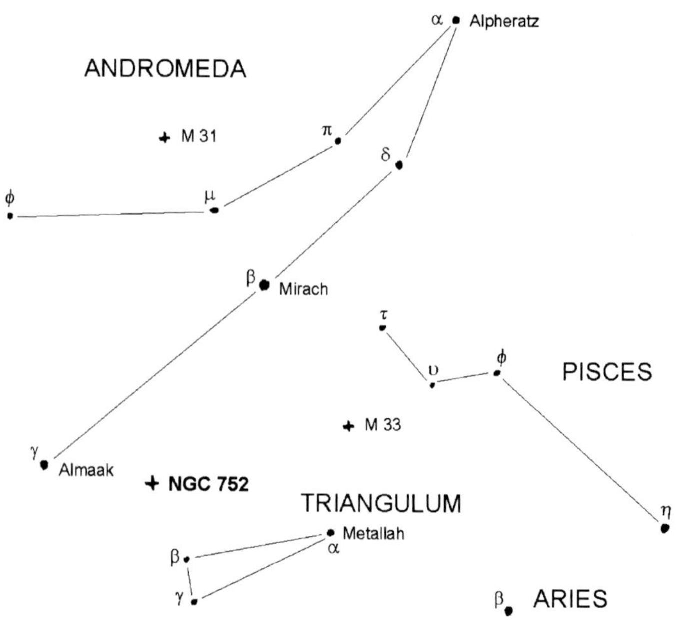

α Alpheratz
ANDROMEDA
M 31
π
δ
ϕ
μ
β Mirach
τ
υ
ϕ
PISCES
M 33
γ Almaak
NGC 752
TRIANGULUM
η
Metallah
α
β
γ
β ARIES

NGC 869 and NGC 884 in Perseus (The Double Cluster)

Type: Open clusters
Designation(s): NGC 869, NGC 884
RA: 02h 20m
Dec: +57°08′
Visual magnitude: 3.9
Distance: 7,300 LY
Size: 60′

The Double Cluster in Perseusis one of the most beautiful clusters in the sky. It has been know for thousands of years and even appears in Hipparcos's catalog of 130 BC. Easily visible in binoculars and amazing in low power telescopes, this pair of clusters is a favorite for fall and winter observers. This object is easily visible naked eye between Perseus and Cassiopeia. Even a small pair of binoculars shows the double knots of stars.

It is somewhat strange that Messier didn't include these objects in his catalog. He included other bright clusters such as the Pleiades (M45) and the Beehive (M44).

H (NGC 869) and Chi Persei (NGC 884), as they are also known, are located only 7,300 light years away and are approaching us at somewhat greater than 20 km/s. Recent findings have shown that although they are located perhaps 300 light years apart and are approximately 70 light years in diameter, they may not have originated together. NGC 884 (5 million years) seems to be a bit older than NGC 869 (3 million years). Many of the surrounding stars also seem to be linked gravitationally to the clusters, perhaps formed within one or both of the clusters and in the process of dissipating. Nonetheless it makes for a striking view at low or high powers.

Incredibly bright supergiant stars, some of which are more than 50,000 times brighter than the Sun, dominate the two clusters. There are both red and blue supergiant stars, making this a very striking cluster both visually and photographically. The maximum lifespan of some of these massive stars are around the age of the clusters; this means at sometime in the near future, one or more may explode as a spectacular nova. But don't hold your breath waiting – it could be thousands of years yet (or tomorrow night!). That's what makes astronomy such a fun endeavor!

Both clusters are similar in size and contain approximately 5,000 solar masses with perhaps 100 members each. However this 5,000 solar masses shines with the luminosity of 200,000 Suns.

Photographically one could not ask for a finer target. The brightness of the pair make nice photographs possible in less than 10 min. With CCD technology, the exposures are greatly reduced.

Suggested Instruments

binoculars
2.5″+ refractor
4″+ reflector
3″+ catadioptric

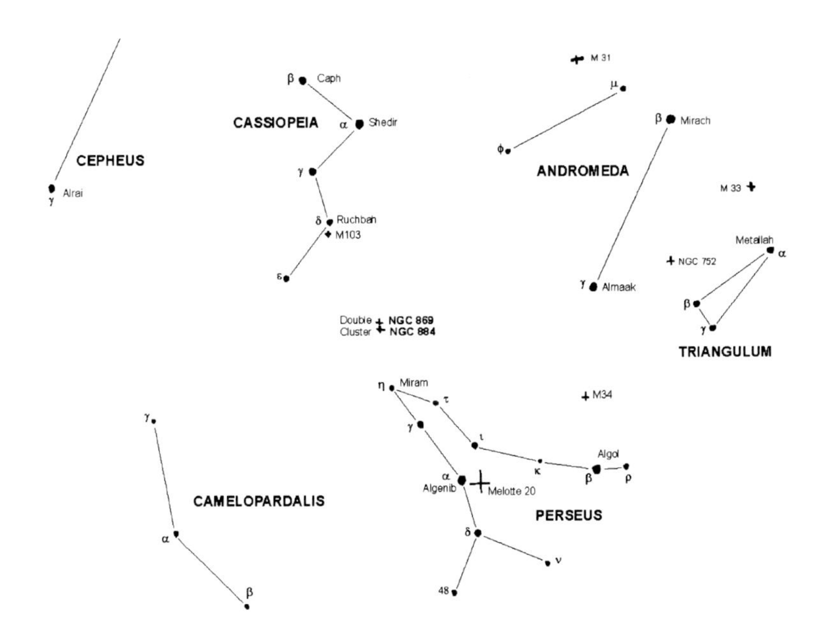

CEPHEUS
Alrai
CASSIOPEIA
Caph
Shedir
Ruchbah
M103
ANDROMEDA
M 31
Mirach
Almaak
M 33
TRIANGULUM
Metallah
NGC 752
Double Cluster
NGC 869
NGC 884
CAMELOPARDALIS
PERSEUS
Miram
M34
Algol
Algenib
Melotte 20
48

M34 in Perseus

Type: Open cluster
Designation(s): Messier 34, NGC 1039
RA: 02h 42m
Dec: +42°47′
Visual magnitude: 5.3
Distance: 1,350 LY
Size: 42′

M34 is one of the most spectacular open clusters in the sky. It is quite bright and can be seen naked eye in very dark skies. In a pair of binoculars it is quite nice. Being approximately the size of the full Moon and fifth magnitude, M34 is a good target for smaller 'scopes. You should see wonderful views of this wispy cluster in telescopes as small as 3″. In 'scopes of 5″ or 6″ the spidery arcs of stars begin to stand out. In even larger telescopes the full extent of this cluster really shines. The size of this cluster makes lower powers of less than 50× the best choice.

M34 is located more than 1,300 light years away and is over 15 light years across, which makes it a truly immense object. With slightly more than 100 members this isn't a very dense cluster. In fact the star density in M34 isn't much more than the Sun's neighborhood. The stars within M34 are much younger (and brighter) than the stars in our neighborhood. M34 is only 120 million years old and still contains a good number of brighter giant stars. Several of these giant starts are white stars of more than 50 times the luminosity of the Sun.

This cluster is exceptional in smaller telescopes under clear skies, sometimes more so than with larger apertures. The larger telescope seems to wash the beauty of the star field with so many additional field stars that the cluster is lost.

Photographically, this is a favorite of imagers everywhere. Images can nicely show the double star h1123.

Suggested Instruments

binoculars
2″+ refractor
3″+ reflector
3″+ catadioptric

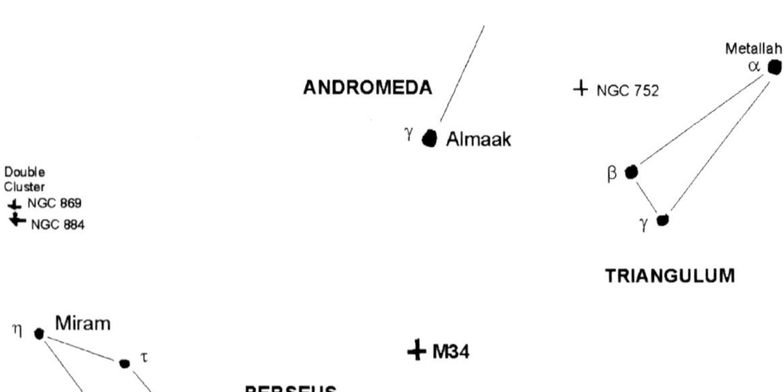
ANDROMEDA
γ Almaak
NGC 752
Metallah
α
β
γ
TRIANGULUM
Double Cluster
NGC 869
NGC 884

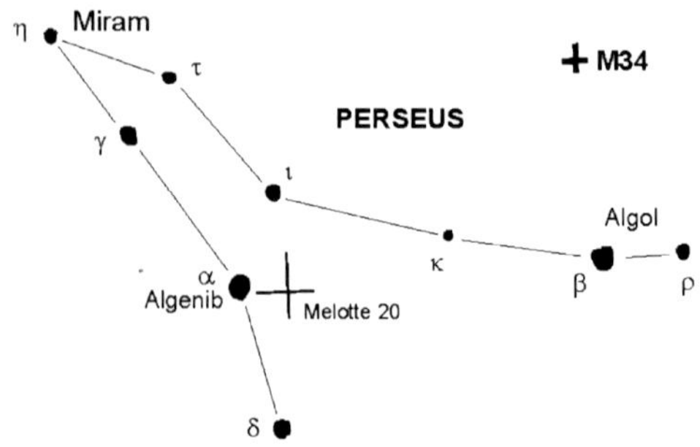
η Miram
τ
γ
PERSEUS
M34
ι
Algol
κ
β
ρ
α
Algenib
Melotte 20
δ

Melotte 20 in Perseus

Type: Open cluster
Designation(s): Mel 20
RA: 03h 22m
Dec: +49°05′
Visual magnitude: 1.2
Distance: 600 LY
Size: 3°

Melotte 20 is also known as the Alpha Persei Moving Cluster. It is located near α (alpha) Persei (Algenib). It contains more than 100 stars and is considered an OB association.

Mel 20 is an interesting object first cataloged by the famous Italian astronomer Giovanni Batista Hodierna and most recently by Melotte in his 1915 list. It is very easy to find and bright, as Algenib, a first magnitude star, is a member.

This beautiful cluster is large and close and has been classified as an OB association. It doesn't seem to contain many O stars but is loaded with bright B and A stars. The O members (the biggest and hottest) all have burned out. When the first generation of bright stars all went nova the expanding pressure waves caused a second generation of star formation, of which the youngest members have also exploded. This wave of nebulosity and star formation is in a region of space known as Gould's Belt. Mel 20 is a product of this region of space. Gould's Belt extends within the galactic spiral arm that the Sun is in.

Photographically, Mel 20 needs a nice 5–10° field to really show its outer regions against the surrounding sky. This is really an object for film photography, although good attempts can be made with wide angle CCD.

Mel 20 is easily located, as the area surrounding and just to the west of Algenib α (alpha) is Persei.

Suggested Instruments

binoculars, finderscope
2″+ wide field refractor
3″+ wide field reflector

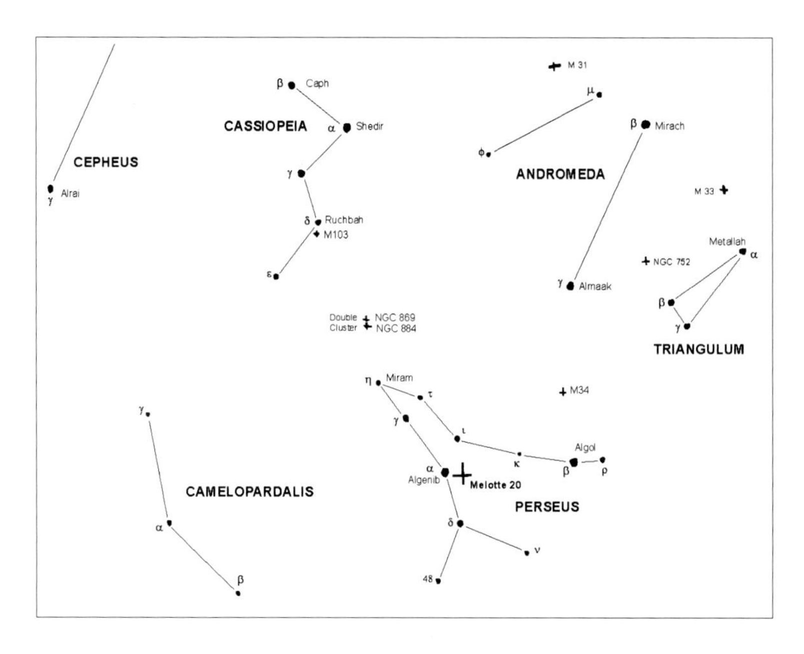
M 31
β Caph
CASSIOPEIA
α Shedir
γ
δ Ruchbah
M103
ε
CEPHEUS
γ Alrai
μ
φ
ANDROMEDA
β Mirach
M 33
Metallah α
NGC 752
γ Almaak
β
γ
TRIANGULUM
Double Cluster NGC 869
NGC 884
η Miram
τ
γ
ι
κ
M34
Algol
β
ρ
α Algenib
Melotte 20
PERSEUS
δ
ν
48
γ
CAMELOPARDALIS
α
β

M45 in Taurus

Type: Open cluster
Designation(s): Messier 45
RA: 03h 47m
Dec: +24°08′
Visual magnitude: 1.6
Distance: 390 LY
Size: 2°

The Pleiades is quite striking in binoculars. It is easy to gaze for prolonged views at this delicious object. In telescopes of 5″ or so you can just begin to see the nebulosity around the brightest stars. In fact the nebula surrounding Merope (23 Tau) is often visible in larger binoculars on a clear, still night. Larger telescopes allow you to see deeper inside the cluster, where more than 300 stars travel through space together.

M45 is more familiarly known as the Pleiades Cluster, or the Seven Sisters. The Pleiades has been known and observed since ancient times. The ancient Egyptians named them Chu, to represent the Goddess Net, "divine mother of heaven." The Japanese call them Subaru, or "those gathered together." (An interesting side note here – if you examine the logo of the car

maker Subaru, you will see the Pleiades stars in the design.) The Hindu call them the flames of Agni, god of fire. The Romans called it the Spring Virgins. The Aztec named them Tianquiztli – "The gathering place." Perhaps the most ancient of all astronomical references is of the Pleiades in an ancient Chinese text of more than 4,000 years ago, naming them Kimah.

The Pleiades was born perhaps 20 million years ago and consists of a variety of young stars. It is located approximately 400 million light years distant and is traveling at some 5.5 arc seconds per century per second to the SSE. This corresponds to 25 miles/s actual space velocity relative to the Sun. They are receding from us at the rate of 4 miles/s.

Pleione, also known as BU TAU, is a variable star of irregular type. It has ejected several shells of gas in the past century and has been studied for X-ray emissions, which seem to indicate it may have a dwarf companion star. For variable star observers, this object is of particular interest, as it represents a certain class of stars called Be-Shell Stars. These are huge stars 3–10 times the diameter of the Sun and many times the Sun's mass.

The Pleiades is an easy object to photograph. From a 50 mm lens on a 35 mm camera to wide field telescopic CCD imaging, the Pleiades is one of the most beautiful objects to image.

Suggested Instruments

binoculars, finderscope
2″+ refractor
3″+ reflector
3″+ catadioptric

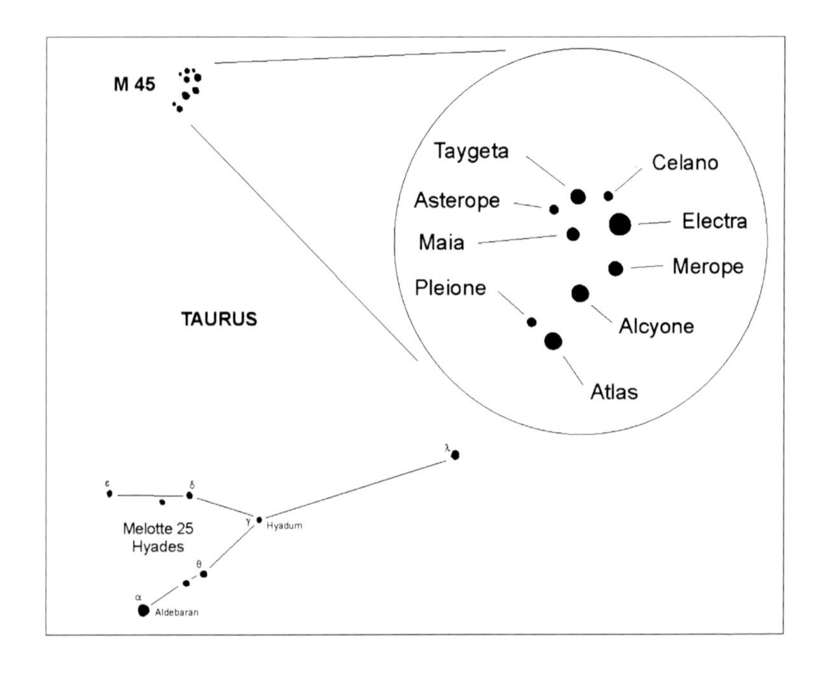
M 45
Taygeta
Celano
Asterope
Electra
Maia
Merope
Pleione
Alcyone
Atlas
TAURUS
λ
ε
δ
γ
Hyadum
Melotte 25
Hyades
θ
α
Aldebaran

The Hyades (or Melotte 25) in Taurus

Type: Open cluster
Designation(s): Mel 24
RA: 4h 25m
Dec: +16°0′
Visual magnitude: +.5
Distance: 150 LY
Size: 5°+

The Hyades are a large open star cluster in Taurus with a distinctive V shape. The V shape is commonly shown on star charts as the face of the bull. Scanning the Hyades you can see a nice variety of colors in the stars. This young nearby cluster is at its best at very low power in binoculars or a wide field telescope.

Very well known by even very ancient cultures, the Hyades cluster is most commonly celebrated as a bull. In many cultures, including the Greeks and Chinese, the Hyades is associated with rain and storms. In classic Greek mythology the Hyades stars were metamorphosed maiden daughters of Hyacinthus, who were sacrificed for the safety of Athens.

At 150 light years this is one of the nearest open star clusters. The central core of the star cluster is more than 10 light years across. However the entire cluster may be as much as 100 light years across and include many other stars in Taurus. The total number of members is difficult to determine exactly due to its large size in the sky. It has been estimated that more than 300 stars are members of this cluster. The Hyades is slowly moving to the east towards a point near Betelgeuse and receding from us as it moves through space at some 25 miles/s.

Although the bright portion of the cluster is 5° in diameter, members are spread out over an area of more than 20°. This gives us a true diameter of the Hyades Cluster of 80 light years. Although this is not a very large cluster, it is of reasonable size and very close. There have been a considerable number of white dwarves and faint stars discovered in the Hyades.

Nearby is the famed variable star T Tauri. This is the prototype of the class of stars known as T Tauri stars. They are a class of young "protostars" that are in the process of final formation. They are stars of similar size to the Sun, but are surrounded by a nebula composed of clouds of gas and dust. In many cases (including T Tauri itself) this surrounding nebula is periodically visible. These stars are quite young, less than a few million years old. They exhibit variability as they gobble up the surrounding nebula.

Suggested Instruments

binoculars, finderscope
2"+ refractor
3"+ reflector
3"+ catadioptric

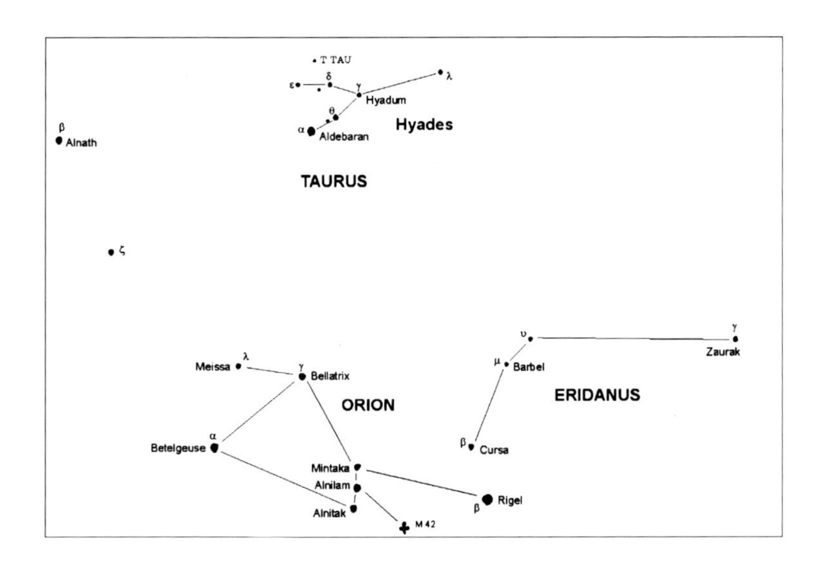
T TAU
ε
δ
γ
λ
Hyadum
θ
α
Aldebaran
Hyades
β
Alnath
TAURUS
ζ
υ
γ
Zaurak
μ
Barbel
λ
Meissa
γ
Bellatrix
ORION
ERIDANUS
α
Betelgeuse
β
Cursa
Mintaka
Alnilam
β
Rigel
Alnitak
M 42

M79 in Lepus

Type: Globular cluster
Designation(s): Messier 79, NGC 1904
RA: 05h 24m
Dec: -24°30′
Visual magnitude: 8.5
Distance: 40,000 LY
Size: 7′

Discovered by Pierre Mechain, who was a friend of Charles Messier in 1780, M79 is one of our more interesting globular clusters. Mechain not only was an avid observational astronomer but was director of the Paris Observatory and did early geodetic work that would later be used to develop the metric system. He discovered numerous astronomical objects during his observations, including 18 objects in Messier's famous catalog.

M79 is interesting because it is believed to be extragalactic in nature. It is located more than 60,000 light years from the center of our galaxy and seems to have come from the Canis Major dwarf galaxy. This galaxy probably had more than a billion stars at one time; however, interaction with the Milky Way has caused it to be considerably disrupted, depositing many of its stars within our galaxy and along its orbit. M79 seems to be one of the numerous shreds that have been torn away in this process. M79 may itself not survive unscathed over the next dozen million years or so.

M79 is also an interesting case as a unique type of star called blue straggler Stars have been found concentrated in its core. These stars are interesting because they are bluer and have more mass than they should for such an old cluster.

Massive blue stars only last for a few million to perhaps a half billion years, and M79 is more than 10 billion years old. So how did these massive stars come to be? It is believed that they evolved either through collisions at the dense core of the cluster or through binary mass transfer, where smaller stars gain mass and appear younger than they really are. These blue stragglers appear Sunken in the core of M79, perhaps making the collision theory more attractive.

These types of objects provide a fixed laboratory in which to study stellar evolution. M79 itself also allows astronomers to study the similarities and differences in stars formed inside and outside of our galaxy.

Photographically M79 is really nice target. The core has a nice rich bluish hue, with the outlying red giants making a striking contrast.

Suggested Instruments

binoculars, finderscope
2″+ refractor
3″+ reflector
3″+ catadioptric

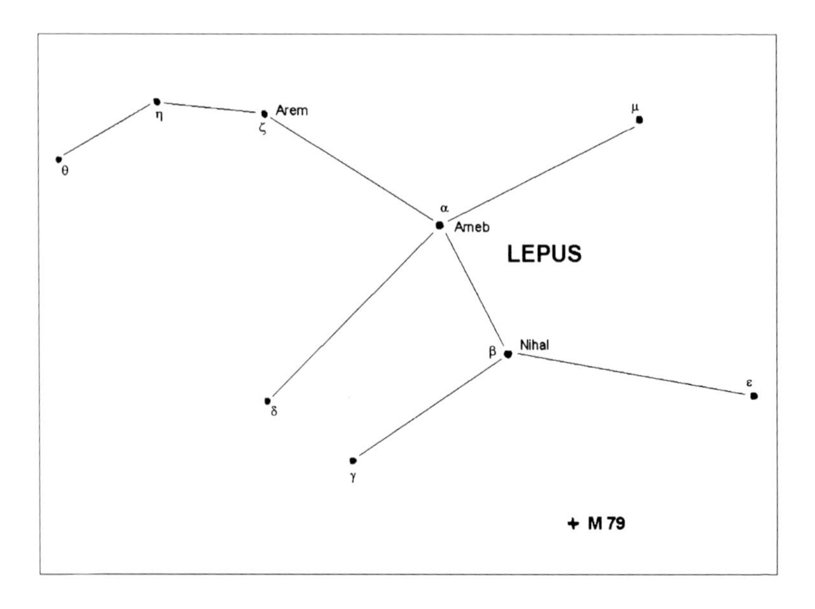
Arem
η
ζ
θ
μ
α
Arneb
LEPUS
β
Nihal
ε
δ
γ
+ M 79

M38 in Auriga

Type: Open cluster
Designation(s): Messier 38, NGC 1912
RA: 05h 28m
Dec: +35°27′
Visual magnitude: 7.4
Distance: 4,100 LY
Size: 20′

M38 is an open star cluster in the center of the Constellation Auriga. This rather beautiful cluster has more than 100 members, with many yellow giants coloring the field.

M38 is one of the finest examples for just enjoying the star field. So many observers rush to see dozens of objects each night without really stopping to enjoy and appreciate what is in the eyepiece. It was more than 4,000 years ago when the light from these stars burst forth on its long journey towards us. At this time the great pyramids at Giza were new, and the Sumerians were at the height of their power in Mesopotamia. Much has happened here, as that light has taken its long, lonely journey to us.

We have learned that planets are not unique, and while M38 is relatively young at 220 million years, we can wonder how many planets might exist around its suns. We can also wonder what it might be like to have a sky full of stars as seen from one of those planets. M38 is more than 25 light years in diameter but has a luminosity of nearly 1,000 Suns. The night sky within this cluster would truly be spectacular.

Images of M38 vary greatly, depending upon exposure. The surrounding star field is stunning with long exposures.

The small cluster NGC 1907 is located about ½° to the south of M38. It is located bit further than M38, but it is older and smaller. Wide field views can encompass both clusters. One interesting point is that these two clusters are only a few hundred light years apart.

Suggested Instruments

binoculars
3″+ refractor
4″+ reflector
4″+ catadioptric

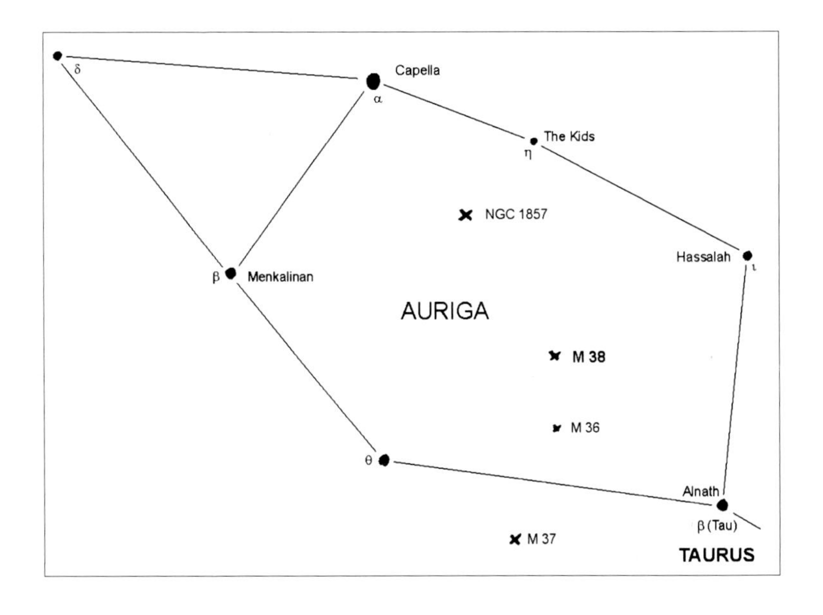
δ
Capella
α
The Kids
η
NGC 1857
Hassalah
ι
β
Menkalinan
AURIGA
M 38
M 36
θ
Alnath
β (Tau)
M 37
TAURUS

M36 in Auriga

Type: Open cluster
Designation(s): Messier 36, NGC 1960
RA: 05h 36m
Dec: +34°07′
Visual magnitude: 5.9
Distance: 4,100 LY
Size: 11′

M36 is a nice bright open star cluster located just 2° from M38. In fact in binoculars, you can easily see both in the same field. M36 is located at about the same distance as M38. It appears brighter because it is much younger (20–30 million years), and it contains many more larger and younger stars that have not yet burned themselves out.

The huge bright stars include one giant star with more than 360 times the luminosity of the Sun. Many of the large bright stars are rotating rapidly. This also occurs in the similarly aged Pleiades star cluster.

M36 has been studied because it has been found to have circumstellar disks around the stars. These studies are undertaken primarily to study planet formation in young stars. When do planets form? At the same time as stars form or subsequently and how? These are all questions that these studies can help astronomers answer.

Imaging M36 is not particularly difficult as it is fairly bright and has a nice condensed concentration of stars at its core. The bright blue and white stars dominate the field. What makes images particularly interesting is the numerous loops and strands of stars.

Suggested Instruments

binoculars, finderscope
2″+ refractor
3″+ reflector
3″+ catadioptric

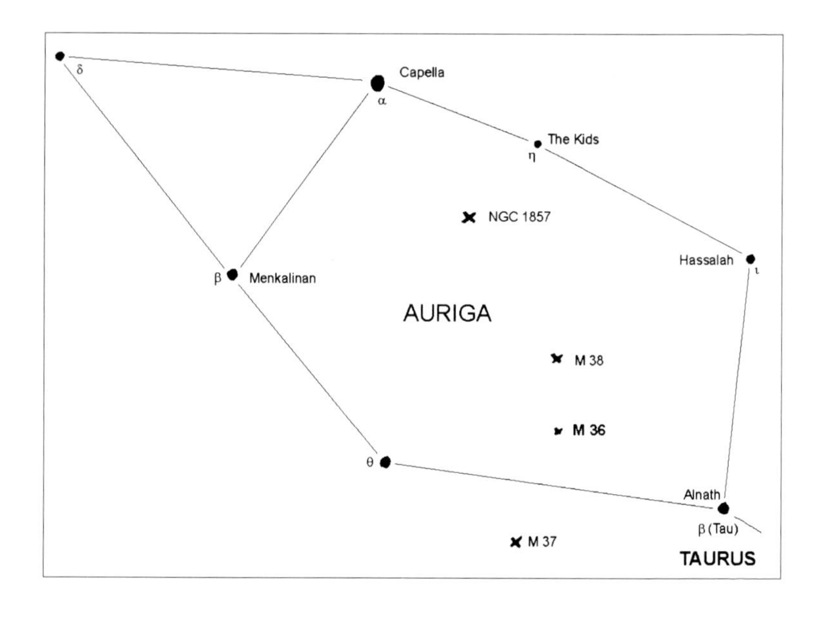
δ
Capella
α
The Kids
η
NGC 1857
Hassalah
ι
β
Menkalinan
AURIGA
M 38
M 36
θ
Alnath
β (Tau)
M 37
TAURUS

M37 in Auriga

Type: Open cluster
Designation(s): Messier 37, NGC 2099
RA: 05h 52m
Dec: +32°33′
Visual magnitude: 5.5
Distance: 4,200 LY
Size: 16′

The brightest of the open clusters in Auriga, M37 is extremely striking and rich. Even in small telescopes M37 reveals a concentrated core and bright outlying stars. When viewed in telescopes of at least 5″ aperture, the stars in the inner core really come to life.

M37 is relatively young at 300 million years old. The most massive stars have already begun to enter the red giant phase, with many bright main sequence stars filling the star field. The cluster is a bit over 25 light years in diameter and contains at least 150 stars.

With a 6″ telescope at 50× this wonderful cluster resolves nicely into dozens of brilliant stars. At times the cluster takes on an almost three-dimensional quality. The lines of stars radiating from the M37 seem to give it an insect-like appearance.

Imaging this cluster can be challenging but is also a lot of fun. Color images are especially interesting, as the age of the cluster makes it have stars of every color from blue, white to yellow orange and red. One of the most prominent stars is a huge brilliant red giant located near the very center of the cluster.

About 2° to the north is the interesting dark nebula B34. In wider field binoculars you can see the noticeable dark area.

Suggested Instruments

binoculars, finderscope
2″+ refractor
3″+ reflector
3″+ catadioptric

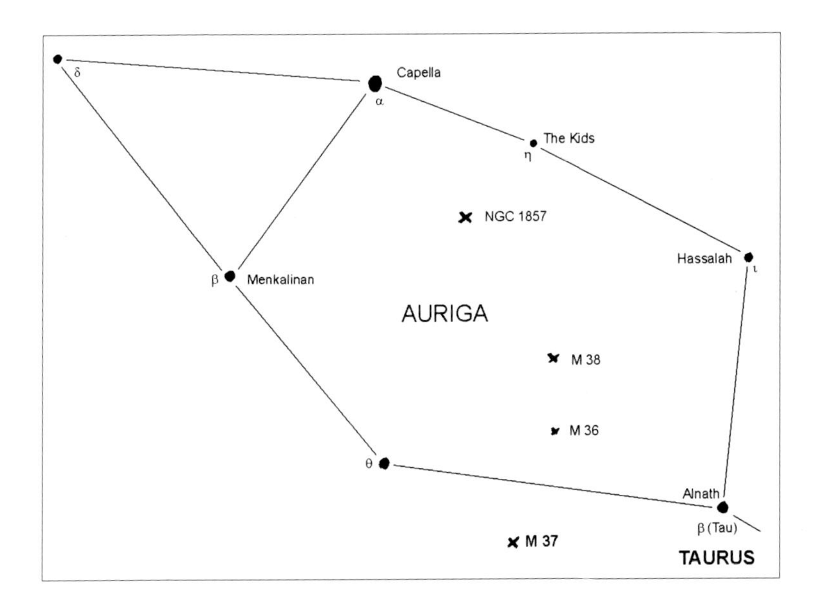
δ
Capella
α
The Kids
η
NGC 1857
Hassalah
ι
β
Menkalinan
AURIGA
M 38
M 36
θ
Alnath
β (Tau)
M 37
TAURUS

M35 in Gemini

Type: Open cluster
Designation(s): Messier 35, NGC 2168
RA: 06h 09m
Dec: +24°21′
Visual magnitude: 5.0
Distance: 2,500 LY
Size: 25′

M35 is a very nice and rich cluster, containing more than 1,000 stars condensed within its 25′size. Only perhaps 200 of these stars are visible in smaller telescopes. The size and distance correspond to a diameter of roughly 25 light years.

The cluster has interesting coloring, due to its intermediate age of around 100 million years. There are some nice yellow and orange dwarves mixed in with the blue and white main sequence stars. This makes for some really exquisite views at different magnifications. It is a good idea with M35 to try changing eyepieces and comparing the views. With certain eyepiece/telescope combinations the cluster seems to rush out at you.

Being somewhat close, M35 has been the subject of several studies involving white dwarves. These studies have helped researchers to understand the proportion of white dwarves in the galaxy.

There is a dark area near the center of the cluster. This region is very noticeable in telescopes of up to around 8″ and can make the cluster appear quite spectacular. It seems to disappear when viewed in larger telescopes except on nights of poor seeing.

The small but rich cluster NGC 2158 (mag. 8.5) is visible just to the southwest in telescopes of at least 6″. There is also another even fainter cluster, IC 2157, just bit further to the west. With wider fields, you can see all three clusters in the same field. The effect is striking, especially with larger apertures.

Imaging M35 is really a lot of fun. You can see it on long exposures even with very wide field views, and higher magnification images really bring out the striking loops and whirls or stars.

Suggested Instruments

binoculars, finderscope
2″+ refractor
3″+ reflector
3″+ catadioptric
Image courtesy of Jan Wisniewski.

(Image courtesy of Jan Wisniewski)

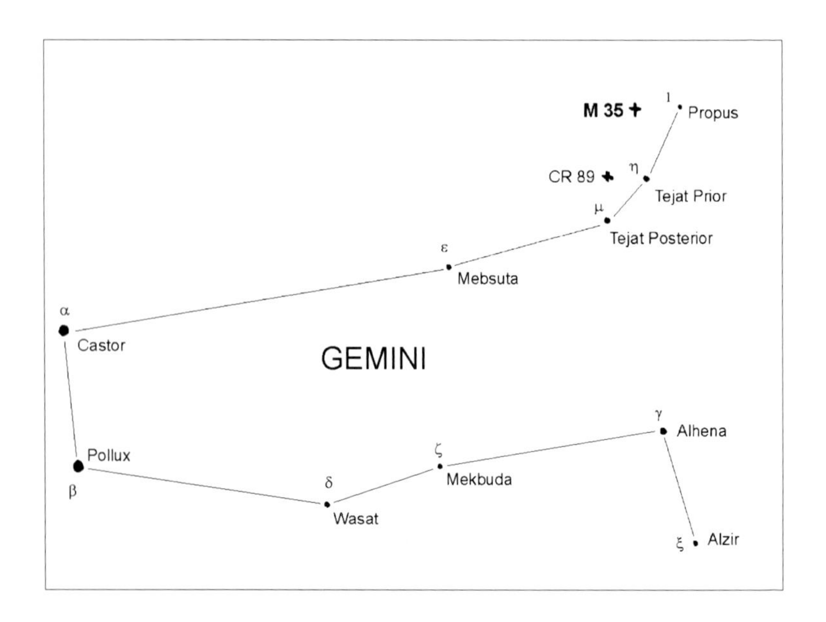

NGC 2244 in Monoceros (Rosette Nebula Cluster)

Type: Open cluster
Designation(s): NGC 2244, NGC 2237
RA: 06h 32m
Dec: +04°52′
Visual magnitude: 4.5
Distance: 4,000 LY
Size: 28′

NGC 2244 is a star cluster in the process of being born. The surrounding nebula, known as the Rosette Nebula, is estimated to have a mass of more than 10,000 Suns. The high energy of the super-massive young stars in the cluster are exciting the gas in the nebula, making it glow. The size of the nebula and cluster is more than 130 light years in diameter. This is one of the largest nebulae known. Several of the stars have masses of more than 30 times that of the Sun.

There are many small dark globules within the nebula. These globule features are seen in other star formation nebulae. These are believed to be dense areas of gas and dust in the process of contracting and have not yet ignited as stars.

The youngest type of newborn stars are known Herbig-Haro objects. These stars often have jets of gas associated with them. These jets have been noticed in NGC 2244, some as large as 3 light years long.

There has also recently been new star formation activity noticed at the edges of the nebula, where the pressure from the energy flowing out from the massive stars in center seem to be pushing the gas and dust together, spurring this newer star development. It is likely that this nebula may spawn more than one cluster before it finally dissipates.

Imaging of NGC 2244 can produce some of the most amazing results. Due to the variety of stars and types of gas and light coming from the nebula, terrific images can be made using various filters, including H-alpha, Oxygen III, red, and others.

Suggested Instruments

binoculars, finderscope
3″+ refractor
4″+ reflector
4″+ catadioptric

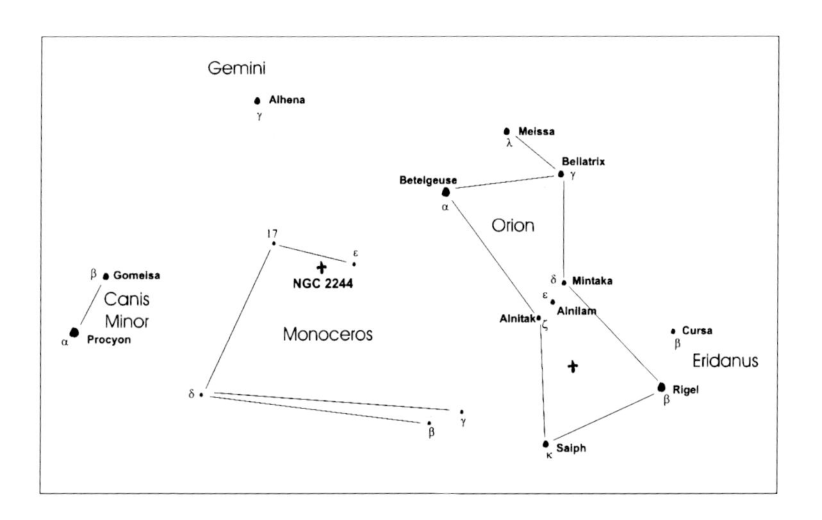
Gemini
Alhena
γ
Meissa
λ
Bellatrix
γ
Betelgeuse
α
Orion
17
ε
NGC 2244
β Gomeisa
Canis
Minor
α Procyon
Monoceros
δ Mintaka
ε
Alnilam
Alnitak ζ
Cursa
β
Eridanus
Rigel
β
δ
γ
β
Saiph
κ

M41 in Canis Major

Type: Open cluster
Designation(s): Messier 41, NGC 2287
RA: 06h 46m
Dec: -20°45′
Visual magnitude: 4.5
Distance: 2,000 LY
Size: 45′

M41 is an excellent object for observing with the naked eye or through modest instruments. It was known by ancient astronomers, even being mentioned by Aristotle in 350 BC. The cluster has more than 200 members with 100+ being visible in modest-sized telescopes. The actual extent of M41 is more than 20 light years. In this cluster is an interesting red giant with a luminosity of more than 700 times that of the Sun. This red star is centrally located, which stands out prominently in even modest-sized telescopes. In telescopes above 8″, this central red star should resolve into two red stars.

At a distance of 2,000+ light years, M41 is quite striking in even small 'scopes. It is receding from us at around 20 miles/s.

M41 is an excellent subject for imaging, as is evidenced by the accompanying images taken by Jan Wisniewski. The bright blue giants make a striking contrast to the central red giant star.

Suggested Instruments

binoculars, finderscope
2″+ refractor
3″+ reflector
3″+ catadioptric

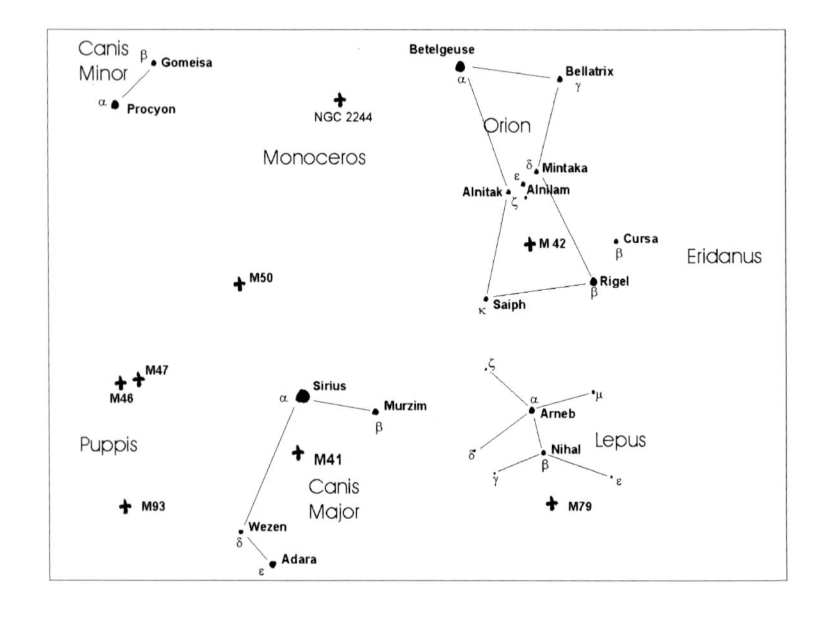
Canis
Minor
β Gomeisa
α Procyon
NGC 2244
Monoceros
Betelgeuse
α
Bellatrix
γ
Orion
δ Mintaka
ε Alnilam
Alnitak ζ
M 42
Cursa
β
Eridanus
M50
Rigel
β
Saiph
κ
M47
M46
Sirius
α
Murzim
β
Puppis
M41
Canis
Major
M93
Wezen
δ
Adara
ε
ζ
α
Arneb
μ
δ
Nihal
β
Lepus
γ
ε
M79

M93 in Puppis

Type: Open cluster
Designation(s): Messier 93, NGC 2447
RA: 07h 44m
Dec: -23°45′
Visual magnitude: 6.1
Distance: 3,400 LY
Size: 23′

M93 is one of the most colorful of the open clusters. It was one of the Messier objects actually discovered by Charles Messier himself. The wedge shape is very noticeable in even a 4″ telescope.

The colors of the stars in M93 range from orange and yellow to blue. The brightest stars in the cluster are in fact blue giants. M93 also has nearly a dozen red giants. These stars have been found to have similar abundance of metals to the Sun. This means that perhaps some of the stars in M93 could have planets similar to those in our Solar System. With the age of M93 being some 200 million years, could there be an early Earth there?

It is really interesting to look at this cluster and wonder if one of the stars we are looking at is an early Sun.

Although the cluster is not very large, it is extremely good for imaging. The bright central wedge of stars and colorful splatter of stars surrounding make for some very striking images. The bright blue stars in the core make a beautiful contrast to the yellow and orange stars in the surrounding area.

The eclipsing variable BU Puppis is located in M93. At maximum BU is only 13th magnitude; however, this is within the range of measurement with 10″+ telescopes.

Suggested Instruments

3″+ refractor
5″+ reflector
5″+ catadioptric

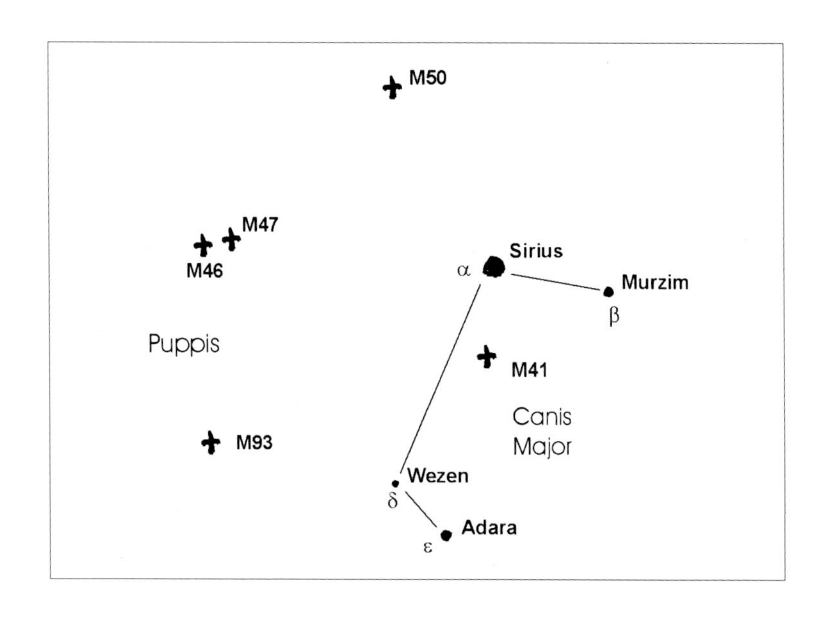
M50
M47
M46
Sirius
α
Murzim
β
Puppis
M41
Canis
Major
M93
Wezen
δ
Adara
ε

M48 in Hydra

Type: Open cluster
Designation(s): Messier 48, NGC 2548
RA: 08h 13m
Dec: -05°47′
Visual magnitude: 5.5
Distance: 1,500 LY
Size: 55′

M48 is a large, relatively close star cluster located in the constellation Hydra (The Sea Serpent). This is a beautiful cluster with brilliant blue and yellow stars. There are over 40 stars down to 11th magnitude, making it a fine object in 3–5″ telescopes at about 50–60×.

There has been some confusion about this object, as Messier placed it about 4° north of where it actually is. This faulty measurement has caused some errors on star charts, including the original Norton's and the seasonal star charts. Messier discovered this cluster in 1777. It is an extremely fine object at lower power. Perhaps Herschel described it best as a "[f]ine large, pretty rich, very straggling cluster of stars ..."

There are several red giant stars in M48 along with a few spectroscopic binaries. This is an interesting cluster about 1,500 light years further out in the galaxy from us. It is a little over 20 light years in diameter and roughly 300 million years old. There are at least 80 members in this fine cluster, many of which are hot blue and white stars. The total luminosity of M48 is estimated to be over 1,000 times that of the Sun.

Photographically, M48 is a wonderful object. With an effective focal length of 1,000–1,500 mm this cluster is beautifully framed with a 35 mm camera and will show detail even with shorter exposures. With longer exposures, the center stands out nicely and shows off the cluster's true variety of colors. With CCD photography, the smaller "plate" size lets you use a shorter lens to get similar results. The beautiful example by Jan Wizniewski on the accompanying page was done with a 135 mm lens.

Located somewhat all by itself in western Hydra, M48 is about 25° south of the Praesepe or Beehive cluster (M44) in Cancer. On some charts it is marked in the wrong location or as a separate object from NGC2548. They are one and the same and at the location specified above.

Suggested Instruments

binoculars 10× and up
finderscope
3–5" wide field refractor
any other telescope that can provide a nice view of about 1° of the sky

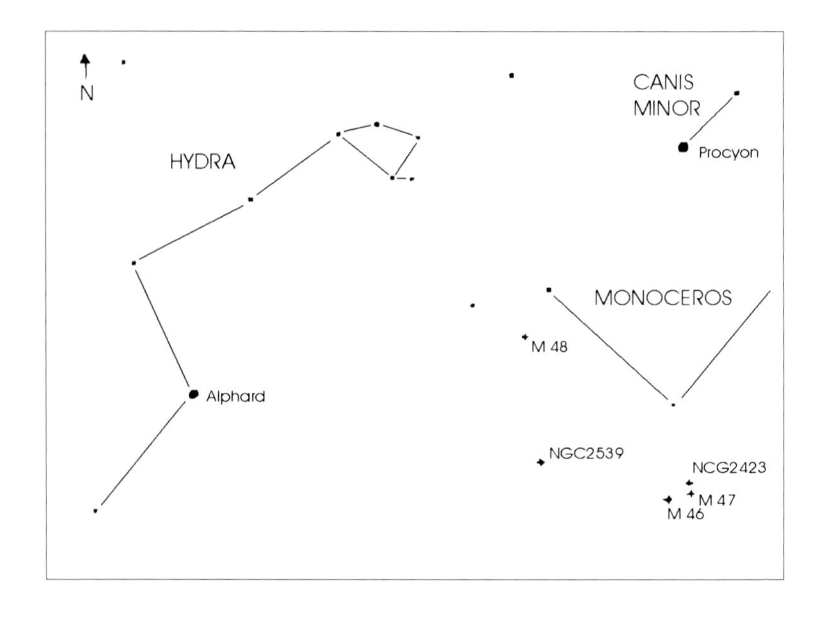
N
CANIS
MINOR
Procyon
HYDRA
MONOCEROS
M 48
Alphard
NGC2539
NCG2423
M 47
M 46

M44 in Cancer (Beehive or Praesepe Cluster)

Type: Open cluster
Designation(s): Messier 44, NGC 2632
RA: 08h 40m
Dec: +19°58′
Visual magnitude: 3.6
Distance: 530 LY
Size: 100′

The "Beehive Cluster" is sometimes called the Praesepe Cluster. This beautiful cluster has been observed since ancient times. There are many ancient references to this cluster, including by Erathosthenes, Ptolemy, and Aratos, the descriptions ranging from "mist" to "cloud." Galileo was the first to discover and resolve the Praesepe into stars, describing it as "... a mass of 40 or more small stars ..." The "Beehive," as it is also known, is easily resolved by smaller telescopes. There are more than 300 stars in the cluster and well over 200 can be seen even in a 4" telescope. The best views can be had with wide field telescopes at lower power.

Many main sequence stars are contained within the Beehive. Several giant stars and white dwarfs are also present. It is believed that this cluster, which

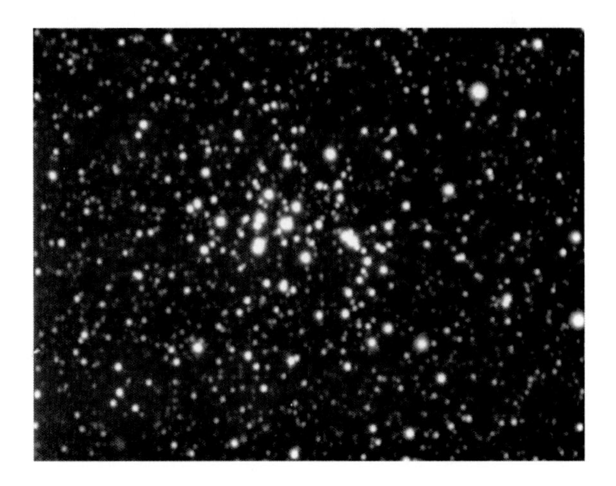

lies about 530 light-years distant, is about 400 million years old. There is also some evidence that this cluster may have been created in the same original nebula as the Hyades cluster in Taurus. The Beehive is about 40 light years across, with a central concentration about 15 lights years in diameter.

The Beehive contains a number of variable stars. TX Cancri is perhaps the most notable variable in the cluster. It is an eclipsing binary with a period of about 9 h. The variable stars BN and BU Cancri are pulsating rotating variables of the δ (Delta) Scuti type.

Delta Scuti stars are fairly rare; however, M44 contains more of these stars than any other cluster. The variables of this class in M44 include BN, BQ, BS, BT, BU, BW, EP Cancri, and KW284.

This class of variable pulsates regularly like stars of the RR Lyrae type, only with much lower amplitude (change in brightness). These stars are A and F class (white) giant stars with a luminosity of 25–50 times that of the Sun and 1.5–2.5 solar masses. Their periods range from about 30 min up to about 5 h.

It is interesting to note that Delta Scuti also seems to share the same motion in space as the Hyades and Beehive cluster, indicating perhaps similar origins.

Photographically, the Beehive is quite spectacular. It is relatively easy to obtain good results, as it has more than 75 stars above the tenth magnitude. Effective focal lengths of 400–800 mm give really nice results. The brilliant blue and white stars in the Beehive make for excellent images.

The Beehive is located right in the middle of the "shell" of Cancer the Crab. It is located only 8° north of another beautiful open cluster, M67.

Suggested Instruments

binoculars 10× and up
finderscope
3–5" wide field refractor

any other telescope that can provide a nice view of about 1° of the sky

The image above was taken by Jan Wisniewski with a Cookbook 245 LDC CCD camera attached to a 135 mm f4 lens on February 14, 1999, from Sooke, BC. It is composed of unfiltered – 10 × 30 s – exposures processed with Multi245, AIP245 and PhotoPaint 8.

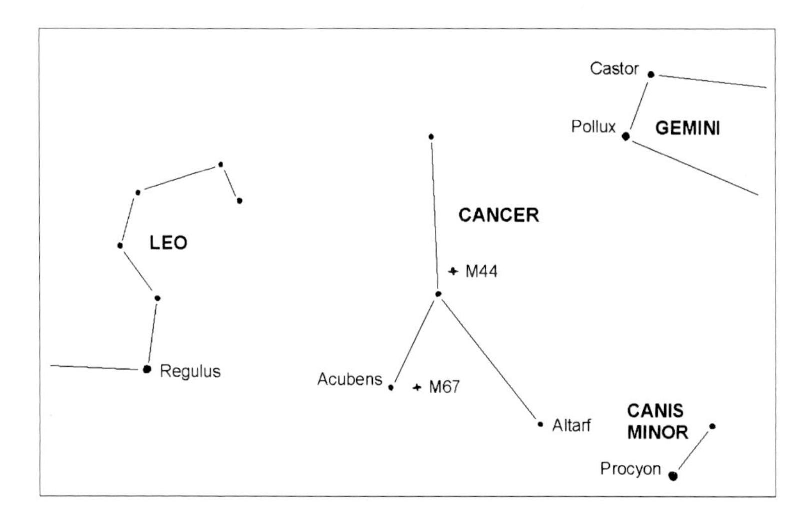

M67 in Cancer

Type: Open cluster
Designation(s): Messier 67, NGC 2682
RA: 08h 50m
Dec: +11°50′
Visual magnitude: 6.1
Distance: 2,500 LY
Size: 30′

M67 is a fine large cluster of contrasting blue and yellow stars. In a telescope of 8″+ this object is quite spectacular. In smaller 'scopes its large size and brightness makes it one of the better clusters for smaller apertures.

M67 was cataloged by Messier in 1780, although it is believed that J. G. Koehler (1745–1801) first discovered this object. Herschel described it as "a pretty rich cluster ..." More than 200 stars can be viewed in telescopes of 8″+. In a 3–5″ 'scopes it can be seen as a faint fuzzy patch.

M67 is an extremely old cluster at over 4 billion years. This cluster has many large stars, which have been aging off the main sequence. Having more

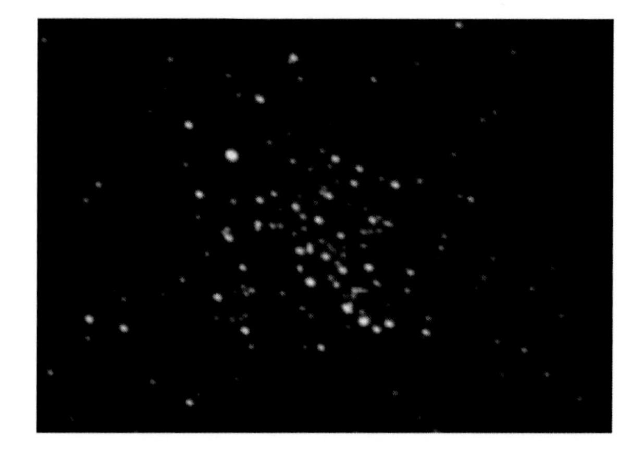

than 500 members in its compact size of about 20 light years makes this an extremely stable object. M67 is also unique in its location. It is found some 1,500 light years above the galactic disk. This is the domain of globular clusters. Even though this cluster is similar in some respects to globular clusters, it is rich in metals like the Sun is. Such metal enrichment makes it plausible that some of the stars in this cluster could have planets. In fact more than 100 Sun-type stars have been identified in M67.

Photographically M67 requires longer exposures and preferably apertures of 8″+ as it is a sixth magnitude object. With CCD or faster film nice results can be obtained. You can obtain some extremely striking pictures with 10–12″'scopes.

M67 is located just south of the "shell" of Cancer the Crab. It is also some 8° south of the large bright open cluster M44.

Suggested Instruments

4″+ refractor
6″+ reflector
5″+ catadioptric

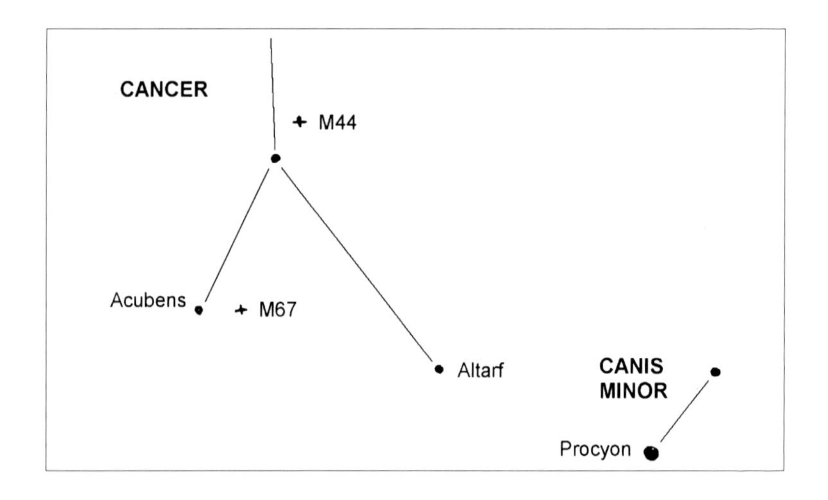
CANCER
M44
Acubens
M67
Altarf
CANIS
MINOR
Procyon

Collinder 285 in URSA Major (URSA Major Moving Cluster)

Type: Open cluster
Designation(s): Collinder 285
RA: 12h
Dec: +55°
Visual magnitude: 2
Distance: 70 LY
Size: 25°

This cluster is the closest known cluster at 75 light years distant. It contains most of the bright stars of the Big Dipper, with a total of about 15–20 members. There are also perhaps 100 stars scattered around the sky, which appear to be lost members of the cluster. These additional stars, including Sirius, share the same motion in space as the cluster. Our own Solar System is located on the fringes of the cluster.

This cluster is obviously a large object, and only individual members may be viewed with a telescope. Even with binoculars the whole extent of it cannot be viewed at once.

There are a few interesting objects in this cluster. The double star Mizar (ζ Zeta Ursa Majoris) is a fine double star separated by 14 arc seconds, making it easily viewed even in a small 'scope. The two stars are magnitude 2.4 and 4.0. The period of rotation of the two stars is estimated to be about 200 years, with a true orbital separation of 36 billion miles. The brighter component (Mizar A) is actually a close binary star, with each component being about 35 times as bright as the Sun and separated by about 27 million miles. Mizar B is also itself a double star, making Mizar a quadruple system. The star Alcor (80 Ursa Majoris) is located only 12 arc minutes from Mizar and makes a nice visual double. Alcor is located about 3 light years behind Mizar, making it not part of the Mizar physical system.

The Big Dipper is always visible for observers north of latitude 40°. It rides high in the sky upside down in the spring.

Suggested Instrument

the naked eye

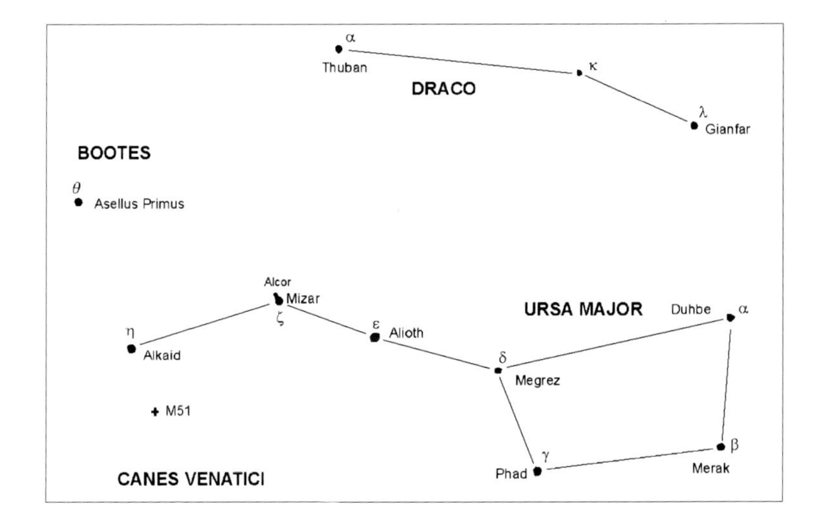
α
Thuban
κ
DRACO
λ
Gianfar
BOOTES
θ
Asellus Primus
Alcor
Mizar
ζ
ε
Alioth
URSA MAJOR
Duhbe
α
η
Alkaid
δ
Megrez
M51
β
γ
Merak
Phad
CANES VENATICI

M68 in Hydra

Type: Globular cluster
Designation(s): M68, NGC 4590
RA: 12h 39m
Dec: -26°45′
Visual magnitude: 7.6
Distance: 33,000 LY
Size: 12′

M68 is a large globular cluster that is nicely viewed in a 4–5″ refractor or 6–8″ reflector. In larger instruments many more stars begin to resolve.

As a very ancient object, M68 is well over 140 light years in diameter. It contains a dense concentration of old stars, with over 250 giant stars. These huge stars have an absolute magnitude of less than 0, which is about half that of super globular clusters M13 and M3. M68 is a large, respectable cluster of hundreds of thousands stars. The overall spectral class is A6 (blue-white), which is fairly typical for globular clusters.

This beautiful object contains over 40 variables, many of which are of the RR Lyrae-type, also known as cluster-type variables. These stars are common in globular clusters and represent large stars that are moving off the main sequence and have begun to pulsate.

M68 lends itself to photographic record with larger telescopes. The use of CCD cameras and "image integration" (the CCD equivalent of long exposure photography) has placed many of the borderline photographic targets such as M68 well into the realm of amateurs. In fact many of the CCD images coming out of amateur hands today rivals serious professional images of a few decades ago.

M68 is located in Hydra about 4° south of the second magnitude star β (Beta) Corvi. About 9° north of M68 is NGC4361 (the Galaxy Nebula), a tenth magnitude planetary nebula in Corvus, so-called because of its resemblance to a galaxy.

Also nearby is the Mira type variable star FI Hydrae. This star is very red, as is typical of long-period variables. It has a period of 324 days and varies up to about ninth magnitude.

Suggested Instruments

4"+ refractor
6"+ reflector
8"+ catadioptric

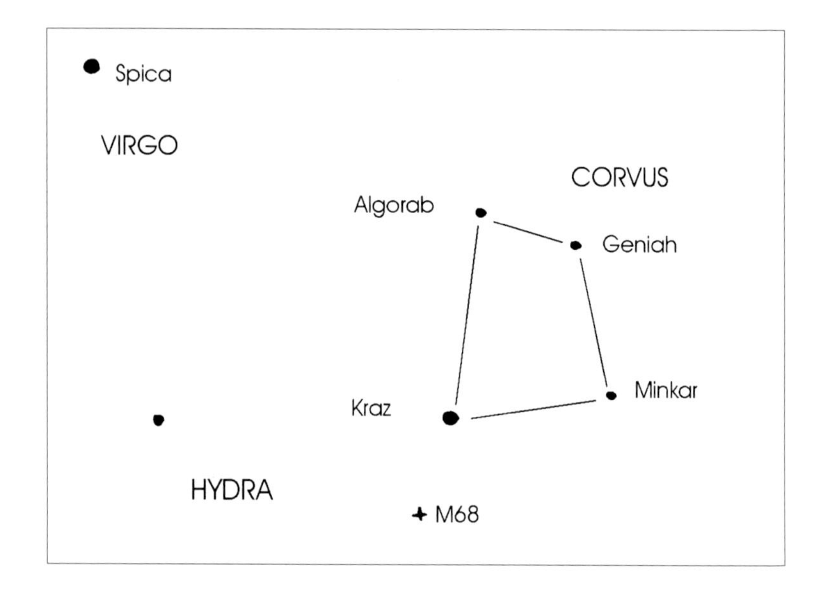
Spica
VIRGO
CORVUS
Algorab
Geniah
Minkar
Kraz
HYDRA
M68

M53 in Coma Berenices

Type: Globular cluster
Designation(s): Messier 53, NGC 5024
RA: 13h 13m
Dec: +18°10′
Visual magnitude: 7.6
Distance: 55,000 LY
Size: 13′

M53 is a distant globular cluster with relatively few stars for a globular. It is a nice object in telescopes from about 4″ and up. In the larger 'scopes many more stars come out. The best observations seem to be with about 50× in a 6″+ telescope. Smaller 'scopes show a small fuzzy spot. In telescopes of 10″ or more the great splendor really begins to come out.

M53 is one of the more distant globular clusters, at more than 60,000 light years from us. The overall luminosity is roughly 200,000 times that of the Sun. It contains over 100,000 stars, over 40 of which have been identified as variables. The variables are of the RR Lyrae (or cluster variables)

type. These variables are similar to the Cepheid variables in that their period is related to their luminosity. This is helpful in determining distances to globular clusters.

This cluster is also known to be particularly "metal poor." Most globular clusters have very little in the way of heavy elements, as they were formed early on in the universe. This cluster must also have been formed early on to be so poor in the heavier elements, which are formed in the hearts of stars and distributed when they go supernova and blast huge quantities of these elements into space.

Less than 1° to the southeast is NGC 5053, a small globular cluster also located about 55,000 light years away. This is estimated to have only around 5,000 stars and a total luminosity of 15,000 Suns. This is very low for a globular. It is about 11th magnitude and a rather difficult object in smaller 'scopes.

Photographically, M53 is quite attractive. With effective focal lengths of 3,000 mm+ it photographs very nicely. Photographs with larger 'scopes and 800–1,000 mm FL can nicely frame both M53 and NGC 5053.

M53 is located about 1° northwest of the Alpha Comae Berenices, a binary star in the eastern part of the constellation. It is also roughly 10° due west of Arcturus.

Suggested Instruments

4″+ refractor
6″+ reflector
8″+ catadioptric

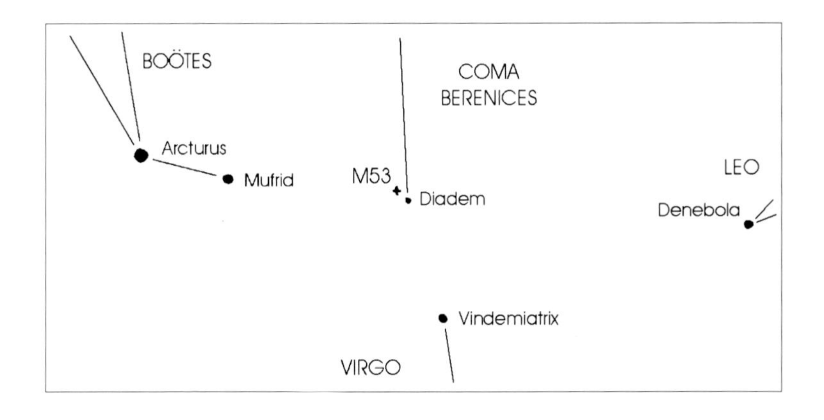
BOÖTES
COMA
BERENICES
Arcturus
Mufrid
M53
Diadem
LEO
Denebola
Vindemiatrix
VIRGO

M3 in Canes Venatici

Type: Globular cluster
Designation(s): Messier 3, NGC 5272
RA: 13h 42m
Dec: +28°23′
Visual magnitude: 6.2
Distance: 34,000 LY
Size: 17′

M3 is one of the most beautiful and rich globular clusters visible. It contains more than half a million stars and is just barely a naked-eye object. Even in small instruments this object is wonderful to observe. In binoculars M3 is a small fuzzy patch. With instruments of 5″ or more, the outer edges show some stars. With an 8″ or 10″ aperture the richness of this cluster is truly amazing. With larger instruments of 12″ aperture or more, this cluster is truly among the best objects visible to the amateur observer. The intense center and sparkling stars extending outward seems to explode into the eyepiece. At 100× this is an object so beautiful, you can spend hours just examining every detail and always finding something new.

With over 500,000 stars M3 is almost a galaxy unto itself. Charles Messier discovered this beautiful cluster in 1764. M3 is over 160 light years in diameter and is more than 300,000 times the luminosity of our Sun. The overall mass of M3 is estimated at over 150,000 that of the Sun. It, like most of the globular clusters in our galaxy, are more than 10 billion years old. The brightest stars visible have already evolved into the giant stage.

This rich cluster has been studied in detail and more than 170 RR Lyrae (cluster) type variable stars have been identified, with more than 200 variables in total discovered. This makes it the number one globular cluster so far for variables in the galaxy. One astronomer alone (S. I. Bailey) identified over 80 of these variables on his own in the 1880s. Bailey discovered these with plates taken with the 13″ Boyden Telescope at the Harvard College Arequipa, Peru, observatory. Many of these variables have such short periods (1/2 day or even less) that they can change brightness noticeably in less than 15 min.

Less than 7° to the east of M3 is the fainter globular cluster NGC 5466. This cluster is located 55,000 light years away and is around ninth magnitude. It is a challenging object for telescopes smaller than 6″. It is somewhat packed with stars; however, a telescope of at least 15″ is required to attempt any semblance of resolution.

M3, like M13, is one of the brightest globular clusters in the sky. It, like most of the globular clusters in our galaxy, are more than 10 billion years old. The brightest stars visible have evolved into giant stars.

M3 is located about midway between Arcturus and Cor Caroli (α {alpha} Canum Venaticorum). It is also located about 10° west of the Coma Star Cluster (Mellotte 111).

Suggested Instruments

binoculars
3″+ refractor
4″+ reflector
4″+ catadioptric

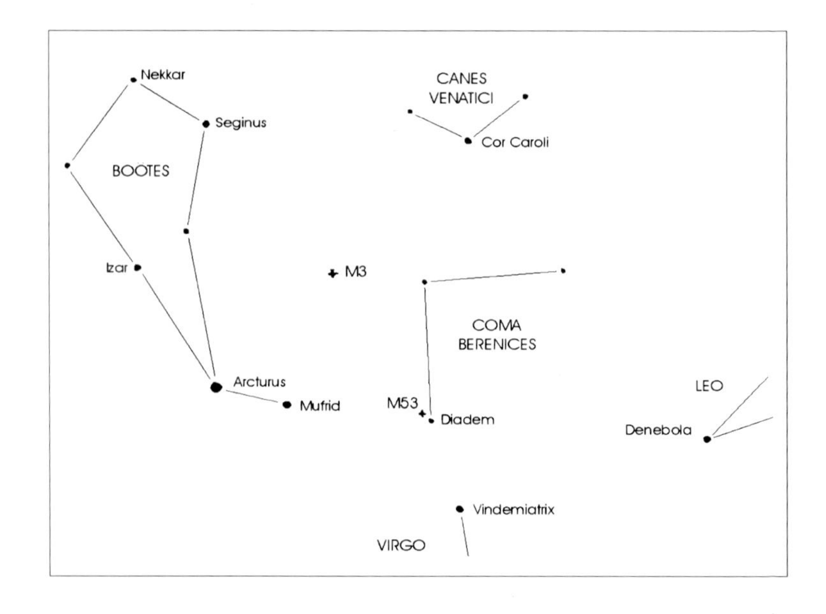
Nekkar
Seginus
BOOTES
Izar
Arcturus
Mufrid
CANES
VENATICI
Cor Caroli
M3
COMA
BERENICES
M53
Diadem
LEO
Denebola
Vindemiatrix
VIRGO

M5 in Serpens

Type: Globular cluster
Designation(s): Messier 5, NGC 5904
RA: 15h 18m
Dec: +02°05′
Visual magnitude: 6.1
Distance: 24,000 LY
Size: 18′

M5 is one of the really spectacular globular clusters in the night sky. This extremely old globular cluster has hundreds of thousands of stars. It is a terrific cluster even in binoculars; although with larger telescopes its true beauty is brought to light.

G. Kirch of the Berlin Observatory first discovered this lovely object in 1702. Kirch was a former student of the famous Polish astronomer Johannes Hevelius. Kirch would later go on to direct the Berlin Observatory and discover a comet and numerous variable stars. The object was added

to Charles Messier's list in 1764, whose small telescope didn't resolve any stars in M5.

M5 is over 150 light years in diameter and contains at least a quarter of a million stars. It is noted for its age of over 13 billion years and its large number of variables. The bulk of these variables are RR Lyrae type variables, as is typical of globular clusters. One variable, however, is a dwarf nova, a unique type of variable star noted for sometimes being the precursors to supernovae.

There are over 100 variable stars in M5, the most noticeable being designated Variable 42 by the Harvard Observatory. M5 changes brightness from 10th to the 12th magnitude. It is located in the outer fringes of the cluster in the southwest.

Just 2° to the south is the 11th magnitude globular cluster Pal-5, located clear on the opposite side of the galaxy from us 75,000 light years away. It is possible to see this cluster in a 6″ telescope as a faint haze. It certainly is a good test of the mid-size telescope. This distant globular was discovered by the Palomar Sky Survey of the 1950s, hence the Pal-5 name. It is believed that it will significantly disrupt the next time it passes near the galactic core.

Photographically, M5 is a terrific object. Its brightness and fairly good size make it easy to get good results. Effective focal lengths of 2,000–3,000 mm prove a nice field of view with 35 mm. Shorter effective focal lengths would be used with CCD imaging, depending upon the chip size. With the nice color variation in this cluster, color photography can create some striking images.

M5 is located in Serpens Caput (The Serpent's Head) about 6° southwest of α (Alpha) Serpentis. It is actually easier to find by looking about 10° due north of β (Beta) Librae. In the finder it is a small fuzzy patch.

Suggested Instruments

binoculars
3″+ refractor
4″+ reflector
4″+ catadioptric

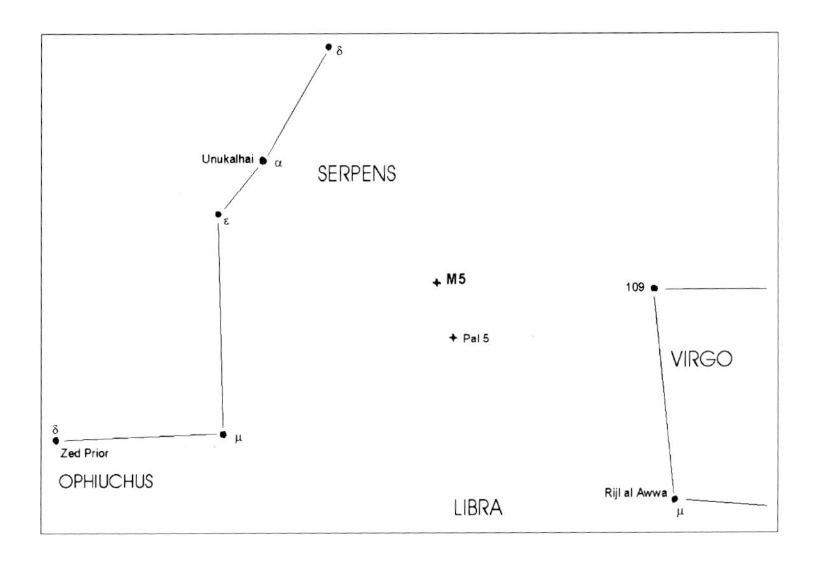
δ
Unukalhai α
SERPENS
ε
M5
109
Pal 5
VIRGO
δ
μ
Zed Prior
OPHIUCHUS
Rijl al Awwa
LIBRA
μ

M80 in Scorpius

Type: Globular cluster
Designation(s): Messier 80, NGC 6093
RA: 16h 18m
Dec: -23°0′
Visual magnitude: 7.2
Distance: 27,000 LY
Size: 9′

M80 is a relatively small globular, best suited for instruments of 8″ or larger. In smaller telescopes it is small and rather indistinct. In large amateur instruments of 10″ and up, it resolves nicely into a small mass with a few scattered stars about it. With telescopes of 13″ and up, M80 really begins to show its nicely knotted outer core of stars.

M80 is home to at least 300,000 stars and is extremely dense. There are indications that this high density has caused many of the stars to lose their outer layers in collisions in the compact core. The information gathered

shows that there are a large number of stellar collisions and close encounters deep inside this beautiful object.

M80 is also the location of at least one known nova, in 1860. As novae are commonly due to interactions between close stars, it is highly probable that additional novae will erupt in this tightly compacted globular cluster. The Hubble Space Telescope has already identified several close novae-type binary systems in M80.

This is what makes viewing M80 and similar clusters so interesting. It is very likely that an amateur observer will discover the next nova seen in M80.

Charles Messier discovered M80 in 1780. He described it as "… resembling the nucleus of a comet." This was quite appropriate, given his instrumentation and task at hand of cataloging objects easily confused with comets.

Just outside the borders of M80 to the east are the two long-period variables R and S Scorpii. They vary in brightness from 9th to 15th magnitude. S varies over a period of 177 days and R of 226 days.

Photographically M80 is best suited to larger instruments. With CCD results can often be quite nice, though, even in 6″ or 8″ telescopes. The small angular size makes eyepiece projection necessary in order to get the image scale to 20 or 30 min diameter for nice framing.

M80 is easy to locate midway between α (Alpha) Scorpii {Antares} and β (Beta) Scorpii {Graffias}. Nearby, just west of Antares, is M4, a beautiful globular cluster.

Suggested Instruments

4″+ refractor
5″+ reflector
5″+ catadioptric

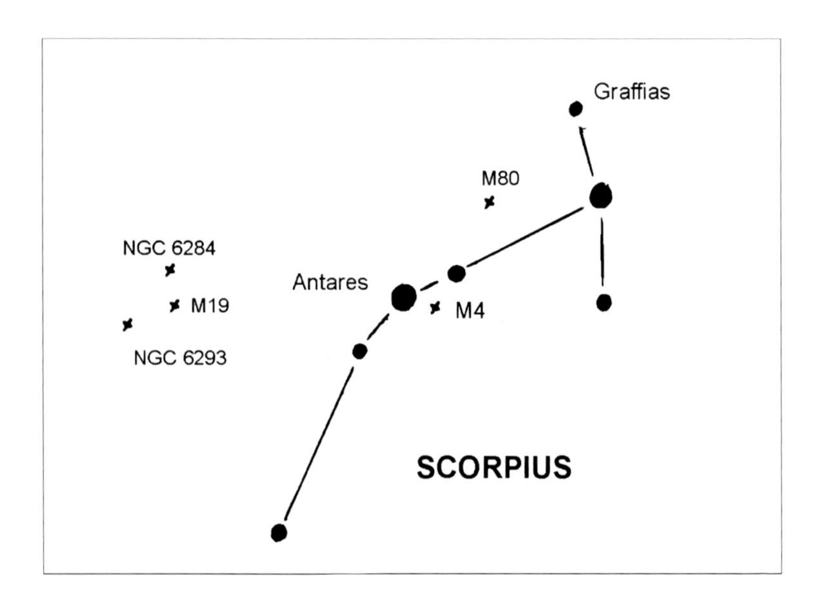
Graffias
M80
NGC 6284
Antares
M19
M4
NGC 6293
SCORPIUS

M4 in Scorpius

Type: Globular cluster
Designation(s): Messier 4, NGC 6121
RA: 16h 24m
Dec: -26°30′
Visual magnitude: 5.5
Distance: 7,000 LY
Size: 27′

M4 is one of the nicest globular clusters in the sky. It is somewhat loosely packed as a globular; however, this makes it easily resolvable into stars even in a 4″ or 5″ telescope. In a 6″ or 8″ telescope it is really a wondrous object. Many nights can be spent looking at M4 through a 6″ telescope, and most viewers can certainly attest to its beauty and almost three-dimensional qualities. Since the extent of this cluster is nearly that of the full Moon, it doesn't require much magnification to see its whole extent. Usually 50× or 60× is sufficient to show it nicely.

Philippe Loys de Cheseaux discovered M4 in 1746. It was later added as one of the first of Messier's famous objects. de Cheseaux was a another

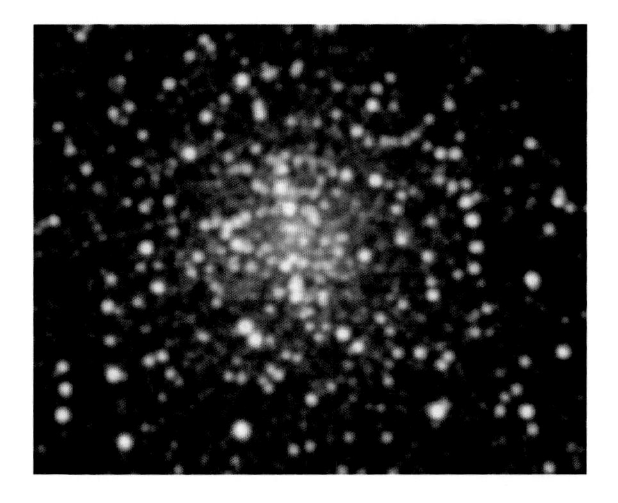

astronomer who, like Messier, compiled a list of nebulous objects. He also discovered M71, another fine cluster in the constellation Sagitta. He is most famous perhaps for his discovery of the famous "six-tailed" comet of 1743. This was one of the objects that Charles Messier observed in his youth and which inspired him to pursue astronomy.

M4 is nearly a naked-eye object, at only 7,000 light years distant. It is also a very sparse globular, with perhaps less than 100,000 stars. It is, however, quite large at over 50 light years across. There is a notable bar-type structure of stars running across the entire cluster. This bar can be seen even in smaller telescopes.

There have been close to 50 variable stars discovered in M4, mostly of the RR Lyrae or cluster variable type. This unique type of star is similar to the pulsating Cepheid type variable star and is useful in determining the distance to objects, as there exist a calculable relationship between the star's period and its luminosity. Interestingly M4 also has pulsar. This particular pulsar is one of the youngest known, with a period of only 3 ms (330 times a second), which makes it even faster than the famous Crab Nebula pulsar.

Photographically M4 is a wonderful object. Its size and brightness make it easy to get good results. The picture that follows was taken using a 500 mm lens. With either CCD or film, M4 presents some interesting opportunities for beautiful pictures. It is also located only about 1.5° from Antares and just under 5° from M80. There have been some stunning 35 mm film pictures taken with 135–200 mm lenses showing the whole area.

M4 is quite easy to locate. It is approximately 1½°s west of Antares. Another fine globular cluster, M80 is located about 4° north.

Suggested Instruments

binoculars
3″+ refractor
4″+ reflector
3.5″+ catadioptric

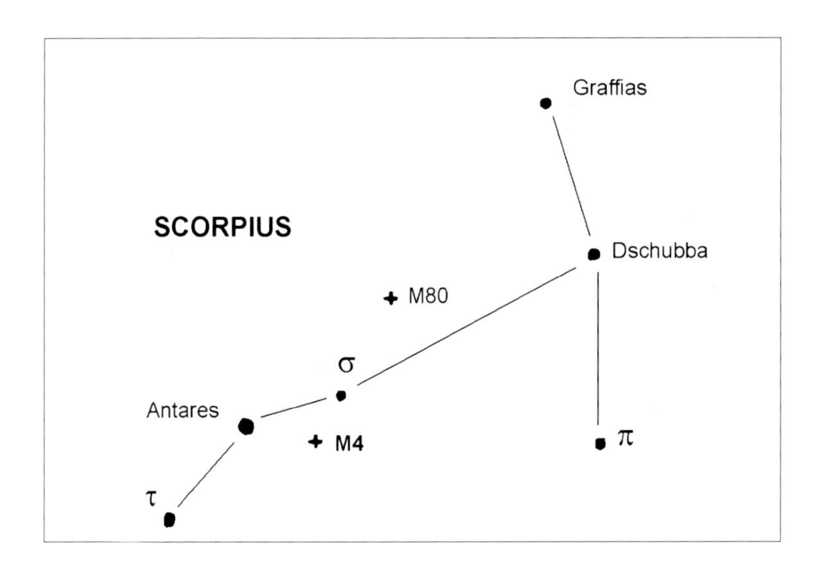
Graffias
SCORPIUS
Dschubba
M80
σ
Antares
M4
π
τ

M107 in Ophiuchus

Type: Globular cluster
Designation(s): NGC 6171
RA: 16h 33m
Dec: -13°02′
Visual magnitude: 7.8
Distance: 20,000 LY
Size: 10′

M107 is one of at was added by his successor, Pierre Mechain, in 1782. M107 is visible as a faint patch in telescopes up to about 7″ or 8″. With larger telescopes the outer stars begin to resolve. This object is a good test of an 8″ telescope for resolving the stars. With 12″ of aperture or more, this is a fine visual object. This cluster presents a nice challenge for the amateur observer with a modest instrument.

The structure of M107 contains some interesting dark nebulae interspersed with the stars. These features are quite striking in an instrument of over 13″ in aperture. They are, however, quite lost in smaller telescopes.

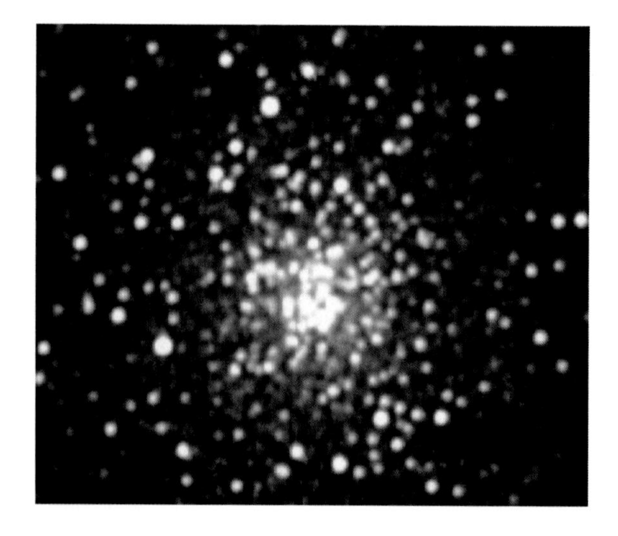

This cluster contains tens of thousands of stars in an area roughly 60 light years across. In contrast there are less than 300 stars in a similar area around the Sun. The stars in M107 are typically metal poor and are quite old, on the order of 10–12 billion years. There are also more than 20 variable stars identified in M107. The bulk of these are the classic "cluster" or RR Lyrae-type variables. These stars pulsate regularly and have been used quite effectively to determine the distances to globular stars clusters. This distant object is approaching us at nearly 150 km/s.

Less than 2° to the northwest is V Ophiuchi, a long-period variable star. This star varies from 7th to 11th magnitude in a bit less than 300 days. These stars are at the end of their lives and are surrounded by clouds of dust, which presumably is being puffed out as the star dramatically changes in size and brightness. The brightness sometimes varies 100 times as the star grows to immense proportions and then shrinks back down again over hundreds of days.

Photographically M107 is challenging, due to its relatively low surface brightness. However with perseverance and a bit of luck very nice results are possible. The photographic plate can reveal much more detail than the naked eye can see. An 8" telescope can produce excellent results, especially with CCD technology.

Located just 3° south-southwest of ζ (Zeta) Ophiuchi, M107 is quite easy to find, as it is a second magnitude blue giant star located about 15° north of Antares.

Suggested Instruments

4"+ refractor
6"+ reflector
6"+ catadioptric

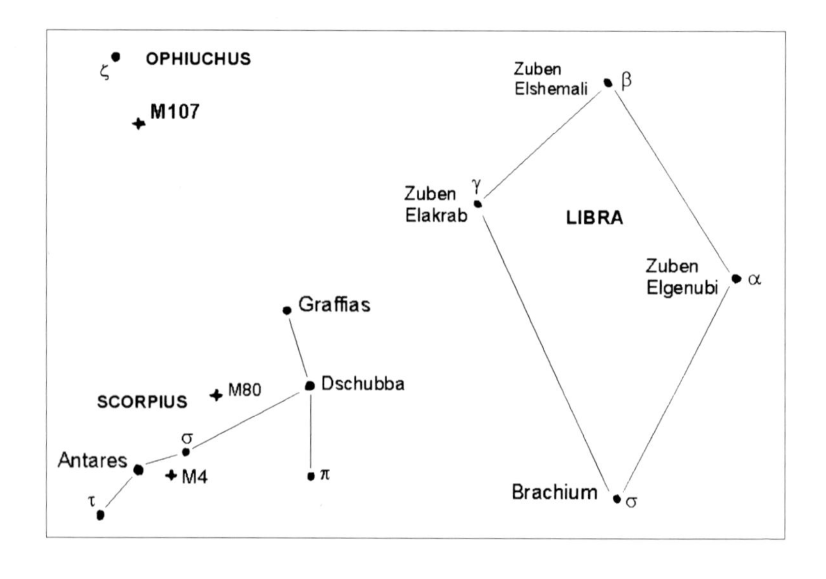
ζ
OPHIUCHUS
M107
Zuben
Elshemali
β
Zuben
Elakrab
γ
LIBRA
Zuben
Elgenubi
α
Graffias
Dschubba
SCORPIUS
M80
σ
Antares
M4
π
τ
Brachium
σ

M13 in Hercules

Type: Globular cluster
Designation(s): Messier 13, NGC 6205
RA: 16h 40m
Dec: +36°28′
Visual magnitude: 5.8
Distance: 23,000 LY
Size: 17′

M13 is perhaps the finest globular cluster in the northern skies. It is a rich and exciting object in virtually all telescopes. The subtle star patterns within it are visible in instruments of 10″ and up. However, in smaller telescopes it can be truly awe-inspiring. This is quite amazing, when one thinks that this object is over 20,000 light years away. This beautiful cluster is also known as the Great Globular Cluster of Hercules – certainly an understatement, as it is surely the finest globular cluster visible in the northern hemisphere.

Discovered in 1714 by Edmund Halley of Halley's Comet fame, M13 has been cataloged by many famous astronomers. Included in this long list of M13 admirers are Charles Messier, who described it as "... round & brilliant, the center is more brilliant than the edges ...." John Herschel described it

as "... hairy-looking ..." in 1833. Such descriptions perhaps were tainted by the lack of instrument quality of the day. However they do indicate that even with inferior instrumentation, some hint of structure is visible.

With more than one million stars M13 is one of the more populous globular clusters in our galaxy. It is well over 100 light years across, with a density of more than one star per cubic light year at the core. It is located about the same distance from the galactic core as the Sun. The age of M13 is well over 10 billion years, perhaps as old as 15 billion years. This makes M13 among the oldest objects in our galaxy.

Interestingly M13 does not contain a large number of RR Lyrae or cluster variables, as do most other globular clusters. Astronomers have been somewhat puzzled by this fact, and much research has been undertaken on this subject.

The faint 11th magnitude galaxy NGC 6207 is located ½° to the northeast of M13. This object is pretty challenging in instruments less than 10″ aperture. It will however show up in wider field photos of M13.

Photographically M13 is quite a nice object. Even with short exposures using piggyback astrophotography, nice results are possible. However owing to its size, pictures through the telescope are preferable. With CCD cameras amazing results are possible. The picture of M13 here by Jan Wisniewski shows how beautiful pictures can be taken with modest-size telescopes yielding results previously only available to professionals with very large instruments.

M13 is easily located in the keystone of Hercules. It is about a third of the way from η(eta) to ζ(zeta) Herculis (the two westernmost stars in the keystone).

Suggested Instruments

binoculars
2.5″+ refractor
3″+ reflector
3″+ catadioptric

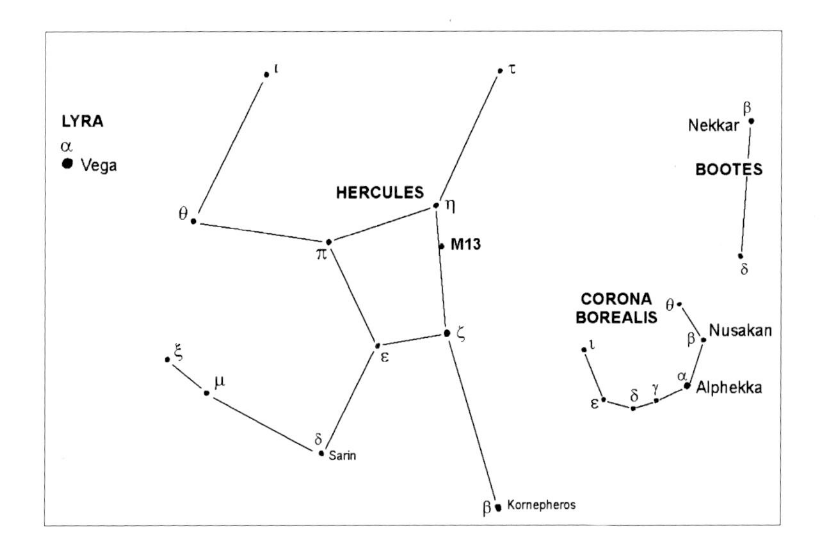
LYRA
α
Vega
HERCULES
M13
ι
τ
θ
π
η
ε
ζ
ξ
μ
δ
Sarin
β
Kornepheros
Nekkar
BOOTES
CORONA
BOREALIS
Nusakan
Alphekka
γ
α

M12 in Ophiuchus

Type: Globular cluster
Designation(s): Messier 6, NGC 6218
RA: 16h 47m
Dec: -01°58′
Visual magnitude: 6.6
Distance: 15,000 LY
Size: 15′

M12 is a beautiful and loose globular cluster that is relatively small and yet quite nice, even in smaller telescopes. With a 3″ telescope this is an easy object to locate. However it does require a larger 'scope of perhaps 10″ to really resolve it nicely. This object seems to lend itself nicely to refracting telescopes. The views through a 6″ or 8″ refractor can be quite as good as a 10″ or bigger reflector. This is perhaps due to its looseness and relative brightness coupled with the superb resolving power of refracting telescopes.

Charles Messier, who discovered it in 1764, described it as round and faint. John Herschel resolved it into stars and described its knots of stars and loose construction.

M12 is located about 15,000 light years away in Ophiuchus and has an actual diameter in excess of 70 light years. In large instruments and in some photographs a hint of spiral structure seems to exist. This may simply be an illusion; however, certain other globular clusters also seem to show this trait. The presence of at least a dozen cluster variables have been noted in M12, which makes it somewhat poor in variables, similar to M13 in Hercules.

An interesting fact is that this cluster seems to be about the same age, size, and distance as its nearby companion globular cluster, M10. In fact they both are perhaps moving around the galaxy together. They are less than 2,000 light years apart, and in each other's sky would appear as large bright objects. They also both have similar and low radial velocities of approximately 15 km/s in approach.

Photographically M12 is more challenging than some other deep sky objects; however, at sixth magnitude, it shows up on even short exposures. The longer exposures are required to really bring out the innermost details.

M12 is quite easy to locate, as it is only 3° from M10 and almost 8½° due east of δ (delta) Ophiuchi.

Suggested Instruments

binoculars
3.5″+ refractor
5″+ reflector
5″+ catadioptric

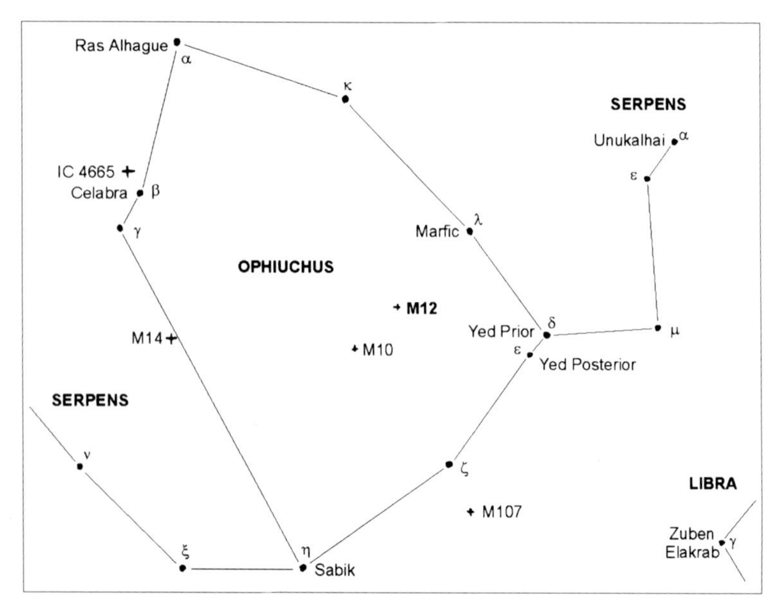
Ras Alhague
α
κ
SERPENS
Unukalhai
α
ε
IC 4665
Celabra
β
γ
Marfic
λ
OPHIUCHUS
M12
Yed Prior
δ
μ
M14
M10
ε
Yed Posterior
SERPENS
ν
ζ
LIBRA
M107
Zuben
Elakrab
γ
ξ
η
Sabik

NGC 6231 in Scorpius

Type: Open cluster
Designation(s): NGC 6231
RA: 16h 54m
Dec: -41°48′
Visual magnitude: 2.6
Distance: 6,000 LY
Size: 15′

A very nice cluster, NGC 6231 is one of nicer objects in the sky. It is quite striking and full of bright, hot young stars. This is a good object for small telescopes. Even a 3″ aperture can really show this cluster nicely. Larger telescopes don't seem to add a whole lot more definition, just brightness. A 5″ telescope is about ideal for this cluster. It's interesting to wonder if the Sun formed inside just such a cluster billions of years ago. Due to its rather southern declination it is a difficult if not impossible object for many northern observers. It is certainly worthwhile to hunt it out when on vacation in southern latitudes.

The astronomer Hodierna discovered NGC6231and included it in his Palermo catalog of 1654. It was not included in Messier's list, presumably because it was too far south from anyplace he normally observed from.

At 3.2 million years old, NGC 6231 is quite young by galactic standards. As with most young clusters, it is dominated by large, hot stars. There have been some recent studies of this cluster showing it to be over 10 light years in diameter.

NGC 6231 also contains a large number of variables. Most notably several β (Beta) Cephei-type variables have been discovered. These unique stars are pulsating stars of small magnitude variations. They may perhaps be precursors to the more well-known δ (delta) Cephei (Cepheid) pulsating stars. There are also several P Cygni variables present. These are extremely massive stars, which periodically erupt with almost nova-like outbursts. They are noted for their unique spectral components. Interestingly enough some quasars also show these strange spectral characteristics.

Photographically NGC 6231 is quite easy. It is well known as a good photographic subject even in smaller instruments. Use a reasonably fast film 400+ ASA for this bright object. Even exposures of 5 min can give good results.

Located at the lower curve of the scorpion NGC 623, it is very close and just north of ζ (Zeta) Scorpii.

Suggested Instruments

binoculars
3″+ refractor
3″+ reflector
3″+ catadioptric

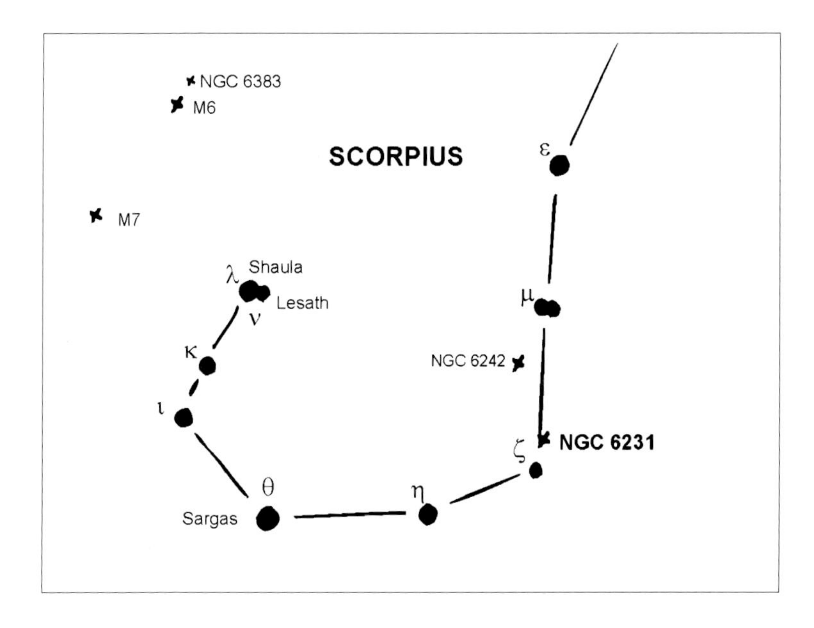
NGC 6383
M6
SCORPIUS
ε
M7
λ Shaula
Lesath
ν
μ
κ
NGC 6242
ι
NGC 6231
ζ
θ
η
Sargas

M10 in Ophiuchus

Type: Globular cluster
Designation(s): Messier 10, NGC 6254
RA: 16h 57m
Dec: -04°06′
Visual magnitude: 6.6
Distance: 14,000 LY
Size: 16′

M10 is a beautiful globular cluster with a rich, dense core. It is an excellent object on telescopes of 5″ and up. With an 8″ or 10″ aperture, the stars begin to resolve and the subtle structure within comes out. Its twin, M12, located only 3° away, is considerably looser, and the pair make a nice contrast. The overall magnitude is 6.6, which makes it visible even in binoculars. However, to really get the best views, a telescope of 5″ or 6″ really is best. With a 12″ or 16″ telescope the soft blue and orange colors of the giant stars abounding is brought out nicely.

This fine cluster has more than 100,000 stars tightly knit together and is perhaps 80 light years across. There are very few variable stars in M10, making

it similar to M13 in that respect. M10 is moving away from us at 70 km/s. It appears to be associated with its nearby neighbor, M12. They are both located about the same distance from us.

A bit over a degree due north of M10 is the long-period variable SS Ophiuchi. This long-period (or Mira-type) variable pulsates from 7th to 14th magnitude in a period of 180 days. These large reddish stars pulsate in and out, growing brighter and dimmer over a reasonably regular period of between 100 and 1,000 days. The first of these stars to be discovered as a variable was Mira (Omicron Ceti) in 1596 by German astronomer David Fabricius who was looking for Mercury at the time.

With the nice dense core M10 is easier photographically than its twin, M12. You can also get some very nice pictures taken with a wider field showing both clusters in the same frame. CCD imaging can provide really fine results, especially as shorter exposures are required.

This cluster is located only 3° from M12 and a bit over 10° southeast of δ (delta) Ophiuchi.

Suggested Instruments

binoculars
3.5″+ refractor
5″+ reflector
5″+ catadioptric

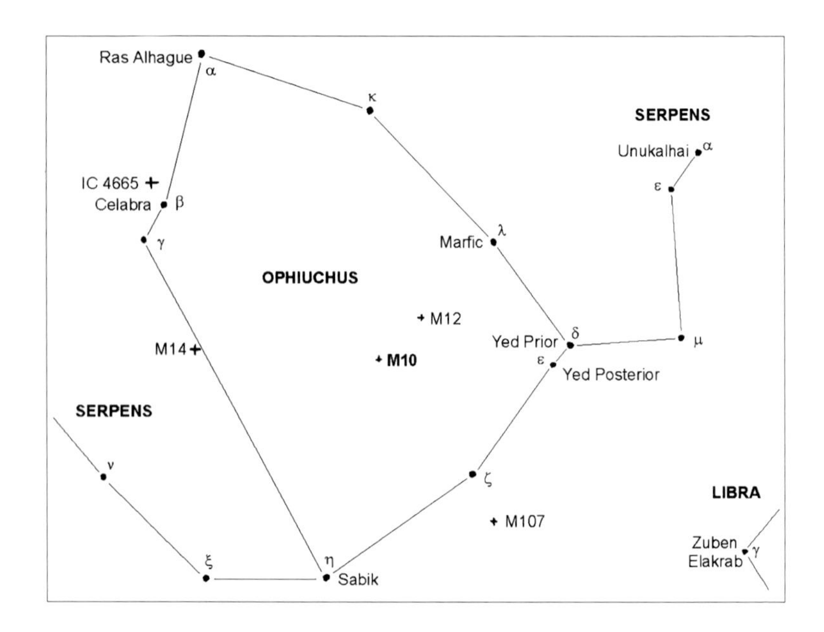
Ras Alhague
α
κ
SERPENS
Unukalhai
α
ε
IC 4665
Celabra
β
γ
λ
Marfic
OPHIUCHUS
M12
δ
Yed Prior
μ
M14
M10
ε
Yed Posterior
SERPENS
ν
ζ
LIBRA
M107
Zuben
Elakrab
γ
ξ
η
Sabik

M62 in Ophiuchus

Type: Globular cluster
Designation(s): Messier 62, NGC 6266
RA: 17h 01m
Dec: -30°06′
Visual magnitude: 6.8
Distance: 22,000 LY
Size: 15′

A misshapen cluster, but certainly intriguing, M62 is bright, dense, and reasonably large. It is a very interesting object in a suitable telescope. A 5″ telescope begins to show some nice details, with views really becoming quite nice in a 10″+ 'scope.

M62, like many of the globular clusters, can be resolved to a much greater degree by long observation. After looking at it for many minutes, new details seem to just appear. You will notice that if you use averted vision and scan back and forth across the field with your eyes, even without

averted vision you begin to notice new and finer details. This perhaps is just an illusion; however, try it sometime!

It appears that M62's proximity to the galactic core has warped it out of shape considerably. This may be one of the things that makes M62 such an interesting object. If you look close enough you can see really fine strings of stars stretching out in whirls and curious shapes near the edges of the cluster.

M62 is one of the objects discovered by Charles Messier himself. He discovered it in 1771 and described it as a "beautiful nebula" and "comet like." Dreyer described it as "remarkable." M62 is located about 21,000 light years away toward the galactic core. It is over 50 light years in diameter, with well over 100,000 stars. Over 90 variables have been identified within M62, with the bulk being cluster (RR Lyrae)-type variables. These stars are older stars that pulsate rapidly as they expand and contract. Their extremely regular light variations are related to their luminosity. This gives astronomers a perfect guidepost for measuring distances to these objects quite accurately.

Interestingly enough, recent work with the ROSAT satellite has shown an X-ray source at the core of M62. This is of particular interest, as visual observations seem to indicate a somewhat recent core collapse of the cluster. This may be an indicator of a large black hole at the center of M62. There is also substantial evidence for neutron star pulsars inside globular cluster cores. A number of gamma ray sources have been found inside their cores, including the core of M62.

Photographically M62 is quite nice. It has a sharp central core and resolves well outside that. This makes photographs quite striking, even with modest exposures. The asymmetry is quite evident in photographs and is just another reason to shoot a few pictures of this really fine and interesting cluster.

Suggested Instruments

4"+ refractor
5"+ reflector
5"+ catadioptric

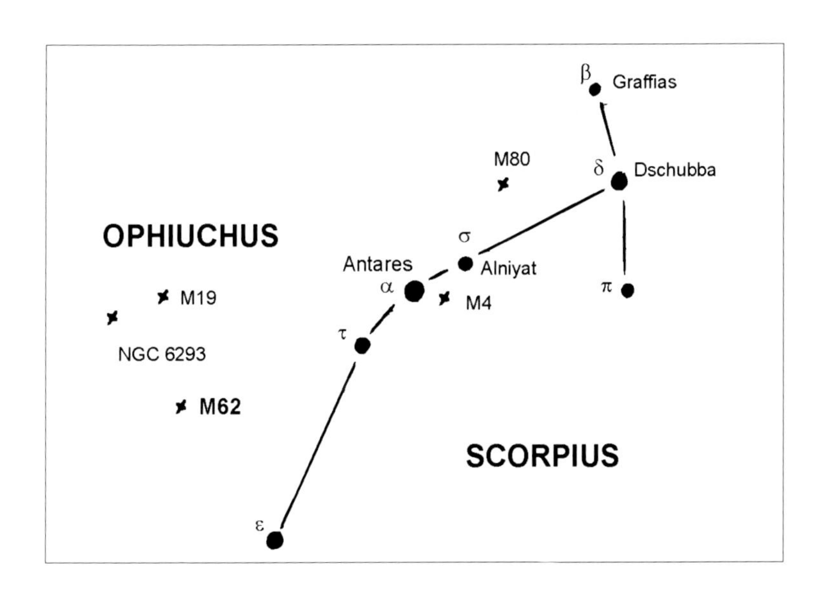
β Graffias
M80
δ Dschubba
OPHIUCHUS
σ
Antares
Alniyat
α
M19
M4
π
τ
NGC 6293
M62
SCORPIUS
ε

M19 in Ophiuchus

Type: Globular cluster
Designation(s): Messier 19, NGC 6273
RA: 17h 02m
Dec: -26°15′
Visual magnitude: 7.1
Distance: 23,000 LY
Size: 14′

M19 is a striking globular cluster located very close to the galactic core. It is known for its oval shape and swirling star patterns. The soft glow of M19 is easy visible in binoculars; however, for really incredible views a telescope is required. There are references stating that M19 really needs a 12″ or larger 'scope to resolve stars in, but in this author's experience this is not true. Perhaps these viewings were made from far northern latitudes, where M19 is on the horizon. In New York, which isn't exactly a southern latitude at 41°N, you can see M19 nicely in a 5″ refractor. The stars on the edges at 150× seem to come alive, and the oval shape is very evident, with little swirls of stars seeming to come off the sides. In an 8″ or a 10″'scope much more detail is available to the observer, and some interesting patterns emerge. The edges seem to be blotchy, as if there are clouds of dark dust surrounding the cluster.

M19 was discovered by the famous comet hunter Charles Messier, who promptly added it to his list of "comet-like" objects in 1764. He described it as a "superb cluster." This is a really fitting description.

M19 is located very close to M62, not only in the sky but in actuality as well. The two clusters seem to be less than 2,000 light years apart and could be much closer. To an observer on a planet in M19, M62 would be many times the size of the full Moon in the sky and very bright indeed.

M19 is over 60 light years in diameter and is receding from us at more than 60 miles/s. It is so close to the galactic core that it has been considerably warped out of shape, like its close neighbor M62.

Unlike its nearby companion, M62, M19 has been found to only have seven variables, all of them RR Lyrae (cluster)-type. These interesting objects are pulsating stars, like the Cepheids, and are very useful in determining cosmic distances.

Less than 1° to the east-southeast is the Cepheid variable BF Ophiuchi. This is one of the extremely regular variable stars used to calculate stellar distances. It is very similar to the RR Lyrae (cluster)-type variables found in globular clusters, only brighter.

Photographically M19 can be pretty spectacular. With its easily resolved areas and interesting shape, it makes for some very striking pictures. CCD images, with their shorter exposures and flexibility, have really moved the amateur a notch up, especially with objects such as M19.

Located only 4° north of M62 and 5° east of Antares, M19 is very easily found. The area is rich with star clouds, and several other faint clusters, including NGC 6293, are located nearby.

Suggested Instruments

4″+ refractor
5″+ reflector
5″+ catadioptric

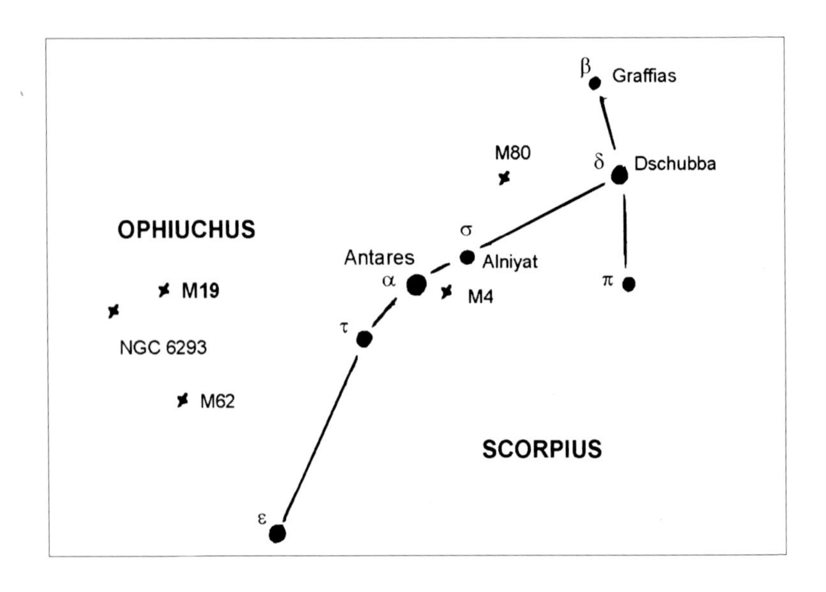
β Graffias
M80
δ Dschubba
OPHIUCHUS
σ
Antares
Alniyat
α
M4
π
M19
NGC 6293
τ
M62
SCORPIUS
ε

M92 in Hercules

Type: Globular cluster
Designation(s): Messier 92, NGC 6341
RA: 17h 18m
Dec: +43°08′
Visual magnitude: 6.3
Distance: 25,000 LY
Size: 12′

M92 is one of the really nice naked-eye globular clusters. If it wasn't overshadowed by nearby M13, it would be the outstanding object of the constellation Hercules.

Observing M92 is an experience even in small binoculars. The cluster is round and bright with stars trailing off at the edges. This appears as a round cottony effect in smaller 'scopes, but in telescopes of 5″ and up, the edges resolve into myriads of stars seeming to flow from the bright central core. This is even an object of beauty in a 3″ or 4″ telescope. The trick is to really study the cluster. Clusters are interesting observationally in that

after you have been looking at them a while, more and more detail seems to come out, sometimes more than you might think possible. This is what makes star clusters perfect objects for amateur observers. Those lucky enough to be able to view M92 through a larger telescope of 15″ or more are in for a real treat.

Located 25,000 light years away and over 80 light years in diameter, M92 is a giant among objects within the galaxy. It easily contains more than 300,000 stars. The age of M92 has been estimated at 15 billion years and the cluster is approaching us at over 100 km/s.

M92 has only been found to have a dozen or so variables of the RR Lyrae (cluster)-type. M92 does have an eclipsing binary variable. This is interesting, as such systems are very rare in globular clusters. The fact that they are rare is due to the fact that close interactions in the central regions of a globular would normally disrupt binary systems.

Just about a degree south of M92 is TX Herculis, an eclipsing binary system with a period of 2 days. There isn't much known about this system, except that it varies in brightness from eighth to ninth magnitude.

Photographically M92 is one of the nicest globular clusters available. Its proximity to M13 makes it a good target for pictures on the same roll. But it is a fine photographic object in its own right. It is extremely beautiful when photographed in color, especially with some of the newer fast films or even a CCD.

Located just north of the keystone, it is pretty easy to find in binoculars or the finderscope. It is located midway between the two stars η (Eta) and ι (Iota) Herculis and is an easy hop north to M92 when at M13. It is certainly worth the hop.

Suggested Instruments

binoculars
3″+ refractor
3″+ reflector
3″+ catadioptric

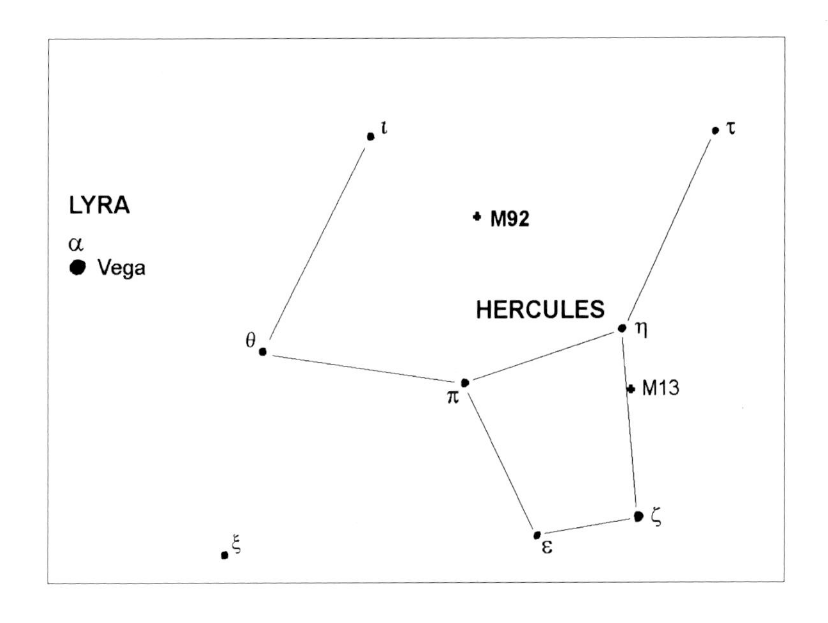
LYRA
α
Vega
ι
τ
M92
HERCULES
θ
η
π
M13
ζ
ε
ξ

M9 in Ophiuchus

Type: Globular cluster
Designation(s): Messier 9, NGC 6333
RA: 17h 19m
Dec: -18°30′
Visual magnitude: 7.6
Distance: 26,000 LY
Size: 10′

Straddled by two vast and dark nebulae, M9 is surely is one of the least dramatic globular clusters known. However, when we realize that some of that dark nebulosity may be obscuring its true splendor, it makes you stop and think. M9 is definitely visible in binoculars. To get a real taste, though, you need a 4″ or 5″ aperture. With a bit of averted vision, you can begin to resolve some of the intricate structure that makes up this unique cluster. With an 8″ or 10″ telescope, even more detail seems to jump out. As you sweep back and forth you can see the dark patches that are Barnard 259 (to the east) and Barnard 64 (to the west). Even though M9 isn't the brightest and most spectacular globular cluster, it has a unique position in being

nestled between two vast clouds of dark nebulosity. In a large enough telescope (perhaps 8″ or 9″), you can clearly see that M9 isn't round, like most globular clusters, but somewhat flattened. Some observers have described it as egg shaped, others as a squished diamond.

Even Messier had an interesting description: "Nebula, without star, in the right leg of Ophiuchus; it is round & its light is faint." Messier discovered this object within the same week as he discovered M10 and M12 nearby. M9 is by far the faintest of this group of clusters.

Located roughly 26,000 light years from us close to the galactic core, M9 is receding at the staggering velocity of 220 km/s. Although this is only a small fraction of the speed of light, it is still an impressive velocity for such a large object. The size of M9 is in excess of 70 light years and it contains more than 100,000 stars.

The dark nebulae nearby are of some interest observationally. They partially obscure the light from M9 and sort of nicely frame it in wider field views. Also just 1° to the northeast is NGC 6359, an eighth magnitude globular cluster.

Photographically, this area is very interesting; with narrow fields you can get nice shots of just M9. With 4° or 5° of image, you can see the dark nebulae, star fields, and even NGC 6359. In the wider field shots, you can even see the edges of the dark nebulae. They look almost like holes in space.

M9 is located about 4° southeast of Sabik (η (eta) Ophiuchi). It is easily found in the finder or binoculars, as it is a bright smudge between two dark patches in the surrounding star fields. Just a degree to the northeast is NGC 6356, a much dimmer globular, just barely visible in larger binoculars.

Suggested Instruments

4″+ refractor
4″+ reflector
4″+ catadioptric

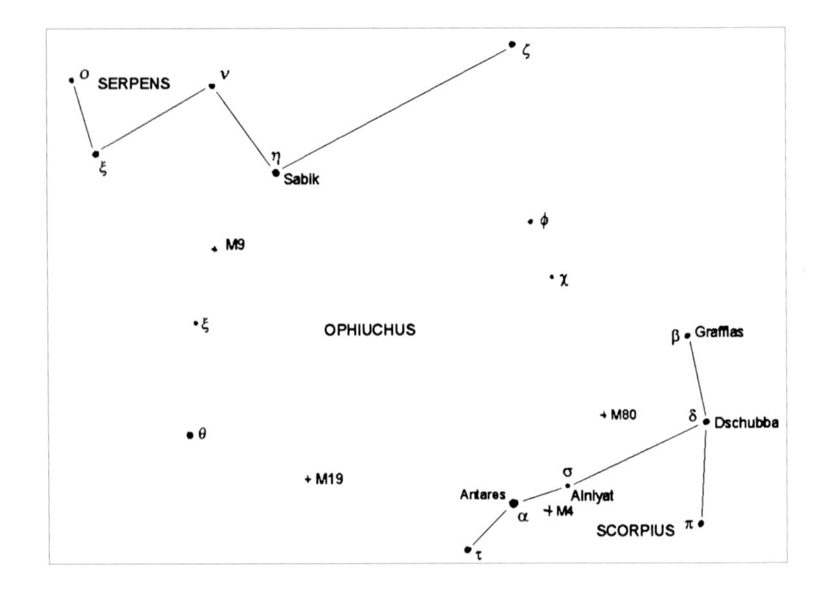

ο
SERPENS
ν
ζ
ξ
η
Sabik
ϕ
M9
χ
ξ
OPHIUCHUS
β
Graffias
M80
δ
Dschubba
θ
σ
M19
Antares
Alniyat
α
M4
SCORPIUS
π
τ

M14 in Ophiuchus

Type: Globular cluster
Designation(s): Messier 14, NGC 6402
RA: 17h 37m
Dec: -03°15′
Visual magnitude: 7.5
Distance: 25,000 LY
Size: 12′

An interesting oval-shaped cluster, M14 is not the brightest globular, but it may be one of the most striking. Its central core is relatively light. This makes it easier to resolve than many of the bigger and brighter globulars. There are some really nice details that can be made out in a 4″ or 5″ telescope. Some nights you can see more detail in a long focus telescope than in faster, larger ones. This may be because, although M14 isn't overly bright, it still is seventh magnitude, and the slower system sharpens the

image up, resolving many more stars. Try M14 out with a mid-sized 'scope and see if you can spot the intricate details. You might even see an almost spiral structure in a 9" telescope, but this may be an illusion.

Charles Messier discovered M14 in 1764, the same week as he discovered M9. He described it as "This nebula is not large, its light is faint." John Herschel, with a much better instrument, saw it as a "delicate globular cluster."

M14 is over 50 light years in diameter and located 25,000 light years distant. It is very rich in variable stars, with close to 100 being cataloged. Most of these are of the RR Lyrae class; however, in 1936, a nova was discovered in M14. The RR Lyrae stars, as previously mentioned, are great distance indicators, as their actual luminosity can be very accurately calculated from their periods.

Photographically, M14 isn't very difficult. It is small, however, and the best results lately seem to be coming from CCD imaging. Resolving stars to the core is possible even with modest equipment and patience. Even traditional photography can sometimes work. The shorter exposure times of CCD makes it a bit easier on the guiding, though.

Located just 6½° south of γ (gamma) Ophiuchi, M14 is pretty easy to spot in binoculars or a small finder. It's pretty well impossible to see with the naked eye, unless you are a cat or superman. Once located it seems to occupy a rather bland section of the sky, and you can pretty easily find it again. It is fairly easy to find γ (gamma) Ophiuchi, and M14 is located just 2° southeast of Celabra or β (beta) Ophiuchi.

Suggested Instruments

4"+ refractor
4"+ reflector
4"+ catadioptric

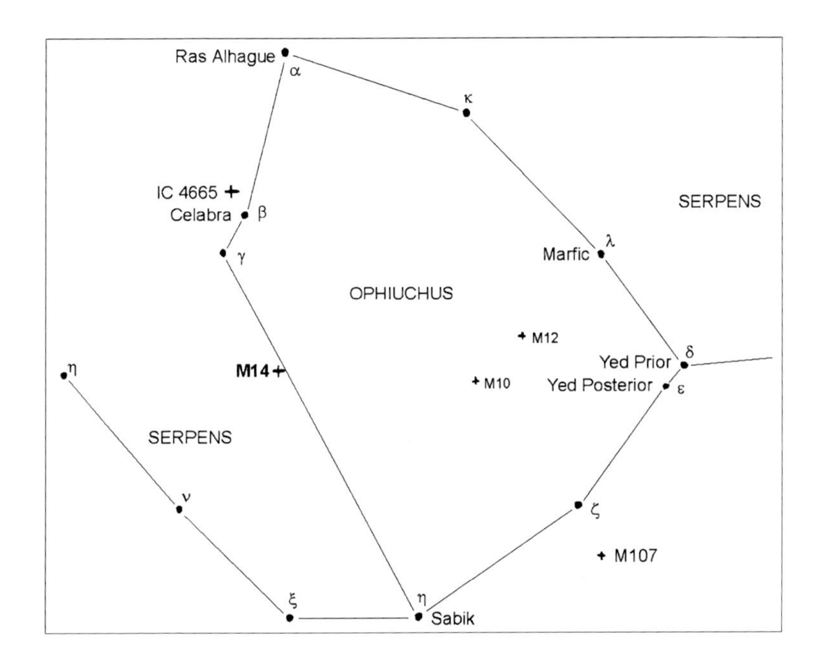
Ras Alhague
α
κ
IC 4665
Celabra
β
γ
SERPENS
Marfic
λ
OPHIUCHUS
M12
δ
Yed Prior
Yed Posterior
ε
η
M14
M10
SERPENS
ν
ζ
M107
ξ
η
Sabik

M6 in Scorpius (Butterfly Cluster)

Type: Open cluster
Designation(s): Messier 6, NGC 6405
RA: 17h 40m
Dec: -32°13′
Visual magnitude: 5.3
Distance: 1,500 LY
Size: 25′

M6 is a rich cluster with nearly 100 stars visible. This cluster is a naked-eye object; however, in a small telescope, dozens of stars resolve. In a 3–4″ telescope you can clearly see most of the members to the core. The single bright yellow-orange giant star contrasts strikingly with the predominantly blue main sequence stars. With a 4–8″ 'scope it resolves wonderfully as one of many people's favorite open clusters. In larger telescopes it becomes something of an explosion of stars that literally fills the eyepiece. The stars seem to align in a delicate butterfly shape, hence its common nickname, the "Butterfly Cluster."

It is believed that this cluster is between 50 and 100 million years old and is located about 1,500 light years toward the galactic center from us. The 50

brightest members vary in magnitude from 6 to 10.5, making all of them quite visible even in small 'scopes. The cluster is perhaps 20 light years across. The brightest star is a K (yellow-orange) giant, with the balance predominantly A and B (blue) main sequence stars.

The second brightest star, HD 160202 (magnitude 6.7), is flare-type variable, which in the past has shown a five magnitude increase in under an hour's time. The last outburst was recorded in 1965. Flare variables are also known as UV Ceti stars. They are usually smaller reddish stars, which periodically exhibit large localized explosions on their surface. They are interesting to monitor because they will increase brightness by several magnitudes within a few minutes time, sometimes even as fast as several seconds, and then go back to normal over 10–30 min.

The brightest star is also known to be a long-period variable (BM Scorpii) with a period of about 800 days. It typically varies between magnitudes 5.5 and 6.5.

Located just 1° to the west of M6 is NGC 6383, a fifth magnitude open cluster and nebula. The nebula is quite faint, but the cluster is very nice. This cluster is somewhat unique in that recently very high energy X-rays have been discovered in a halo surrounding the cluster.

M6 is a very nice object for photography. There have been some amazing photographs with 3″-class telescopes. Of course, exposures of 15+ min will be required with most telescopes to get good results.

Located deep in the heart of the Milky Way in Scorpius, M6 is located a few degrees northwest of another fine cluster (M7). It can be found just north of the scorpion's tail or directly west of the spout of Sagittarius's teapot.

Suggested Instruments

3″+ refractor
6″+ reflector
3″+ catadioptric

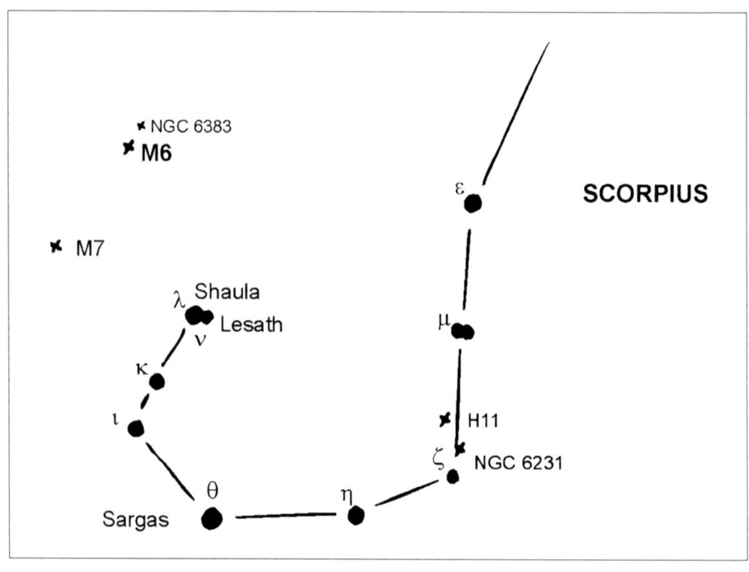
NGC 6383
M6
M7
SCORPIUS
ε
μ
λ
Shaula
Lesath
ν
κ
ι
H11
ζ
NGC 6231
θ
η
Sargas

M7 in Scorpius (Ptolemy'S Cluster)

Type: Open cluster
Designation(s): Messier 7, NGC 6475
RA: 17h 53m
Dec: -34°49′
Visual magnitude: 4.1
Distance: 900 LY
Size: 82′

M7 is a large, bright cluster located only 3½° from M6. It is, however, much closer and brighter than the latter, with many more sixth and seventh magnitude stars. The cluster is also quite large, owing to its relative proximity. At over 1° in size it is one of the larger open clusters visible and certainly the most noticeable deep sky object in Scorpius.

M7 is an ideal object for binoculars or small telescopes. The views of its bright blue stars are quite exquisite, especially in wide field telescopes. Viewed through a 3″ or 4″ F6 refracting 'scope at low power, it seems to jump out at you. This is one object that seems to lose something with high magnification in larger telescopes.

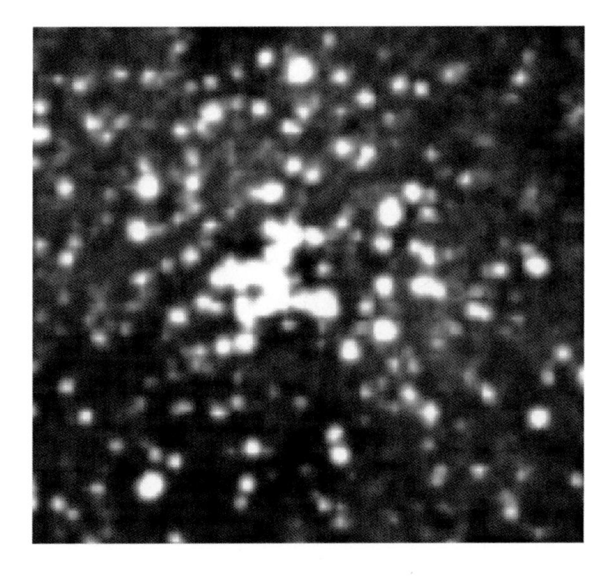

M7 is estimated to be around 250 million years old and is located about 900 light years distant. It contains over 100 stars, many of which are very bright A- and B-type stars. The brightest star is a G8 (yellow) star of magnitude 5.6.

There is a faint globular cluster, NGC-6453, in the field that is an 11th magnitude fuzzy spot, which in larger telescopes resolves as having a brighter center.

With its bright stars and large size, M7 is ideal for astrophotography. With a field of 3–5° it sparkles on pictures. If the field is enlarged slightly and allowed to include M6, some really nice pictures are possible.

M7 is located above the tail in Scorpius about 3½° southeast of the fine open cluster M6.

Suggested Instruments

binoculars, finderscope
3–5″ wide field refractor
any other telescope that can provide a nice view of 3+° of the sky

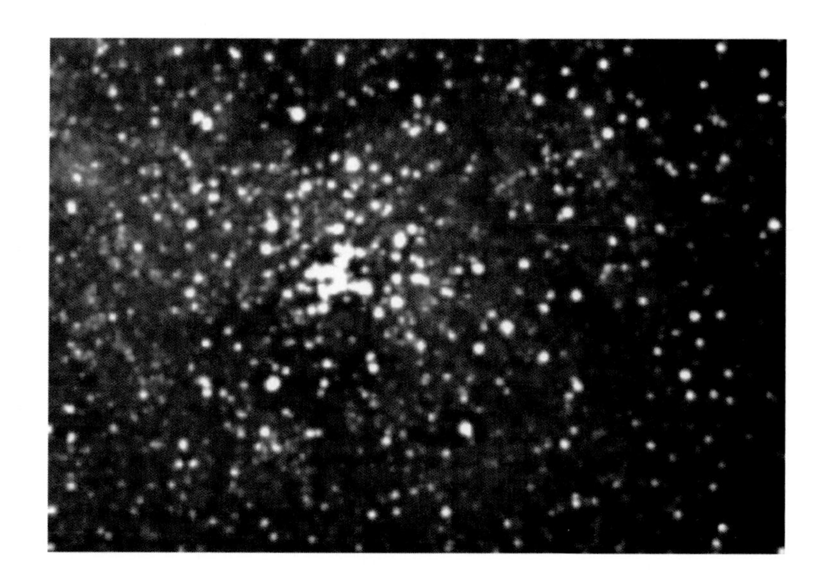

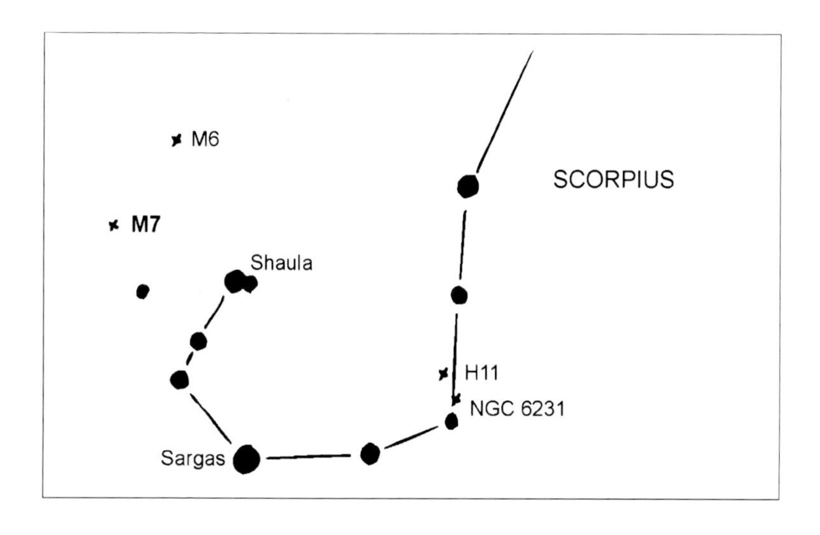
M6
SCORPIUS
M7
Shaula
H11
NGC 6231
Sargas

M23 in Sagittarius

Type: Open cluster
Designation(s): Messier 23, NGC 6494
RA: 17h 57m
Dec: -19°01′
Visual magnitude: 6.9
Distance: 2,000 LY
Size: 30′

M23 is truly a beautiful star cluster. It is roughly the size of the full Moon and contains over 50 bright stars. This object is beautiful even in binoculars; however, in a 3–5″ 'scope at about 50× or 60× it is truly a wondrous sight. In a 12″+ telescope over 100 stars resolve nicely. With its beautiful blue and yellow giant stars this cluster can be observed on many a summer evening.

M23 has over 150 members, some of which are beginning to begin to redden as the larger ones expand off the main sequence. Most of the stars are B (blue) type stars. This stunning star cluster is about 2,000 light years away and is about 250–300 million years old. It has an overall diameter of about 15 light years and has density of about 30 stars per cubic parsec. This makes it a relatively stable cluster. M23 is receding from us at about 8.5 km/s.

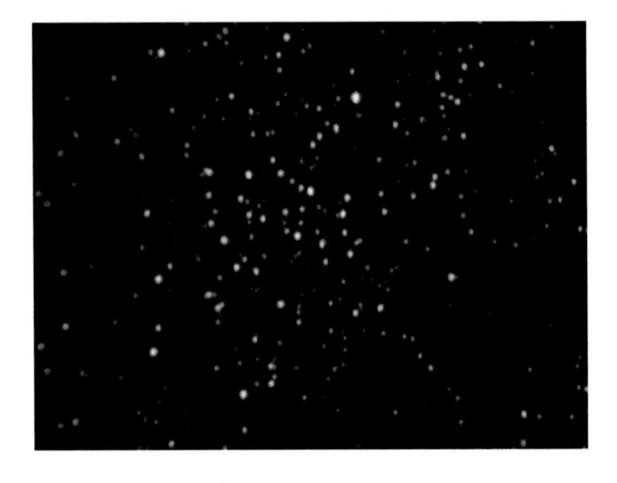

The two brightest stars, along with many of the other members, have been identified as variable stars. The brightest star is a sixth magnitude B1 (blue) giant variable star. The second brightest is an eighth magnitude G8 variable. This gives this cluster a unique personality, which can change a little depending on when you observe it.

Photographically this is a very pleasing object. It can be photographed nicely at prime focus with 1,000–1,500 mm focal length. With larger 'scopes and longer exposures or with CCD technology extremely nice results can be had.

M23 is located in Sagittarius about 12° due north of Sagittarius's teapot. It is only about 4° north of the nice grouping of M8 (Lagoon Nebula), M20 (Trifid Nebula), and M21 Open Cluster. It is also about 5° west of another fine grouping of objects; Clusters M24, M18, and M17 (Horseshoe Nebula). The fine seventh magnitude globular cluster M9 is also located due west of M23.

Suggested Instruments

3″+ refractor
6″+ reflector
3″+ catadioptric

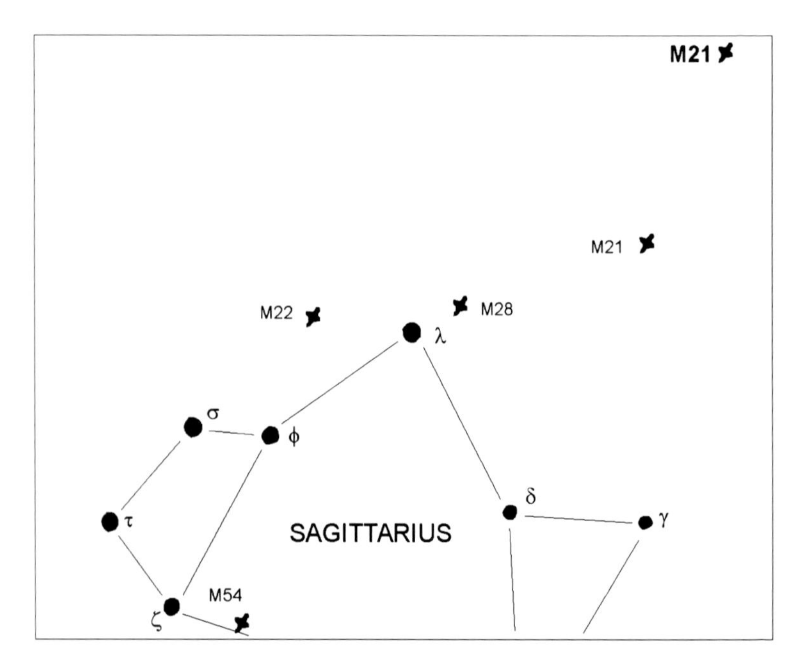
M21
M21
M22
M28
λ
σ
ϕ
δ
γ
τ
SAGITTARIUS
M54
ζ

M21 in Sagittarius

Type: Open cluster
Designation(s): Messier 21, NGC 6531
RA: 18h 04m
Dec: -22°30′
Visual magnitude: 6.5
Distance: 3,000 LY
Size: 15′

M21 is a cluster with a strongly concentrated central region. At magnitude 6.5 it is almost naked-eye visible and can easily be seen in a pair of small binoculars. However, with about 30–50× in a 3″ or 4″ telescope it is quite dazzling; note especially its outlying stars and dense core. It is buried in a region of the Sagittarius Milky Way and is quite close to the Lagoon Nebula. In a larger 'scope the core opens up to reveal many more stars. You might like to view this object at lower power (around 30×), which brings both M21 and M20 (The Trifid Nebula) in the same field. With even lower power (about 15–18×) you can usually get M21, M20, and M8 (the Lagoon Nebula) all in the field.

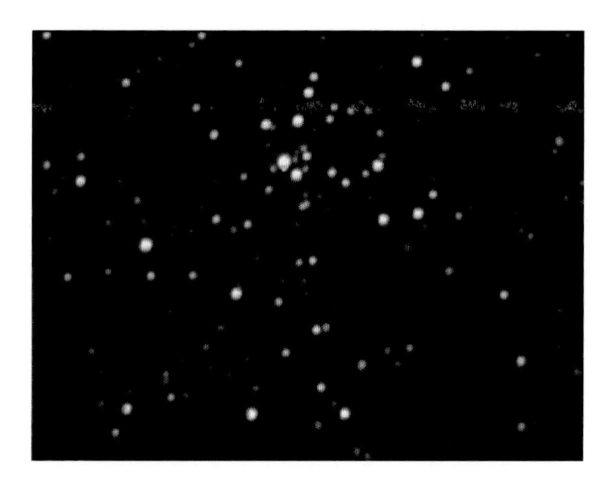

M21 is a cluster whose age has been estimated at only 4 million years, a relative newcomer to the galaxy. The stars are mostly hot B-type (blue) stars. The cluster is probably about 15–20 light years in diameter, making the central density about ten stars per cubic parsec. This is relatively dense for an open cluster. The estimates for its distance vary quite a bit, from 2,000 to 4,000 light years, with 3,000 seeming to be the best guess.

Photographic results are quite good, as there are a number of surrounding objects of interest. Larger fields of 3° or 4° will reveal M8, M21, and M22. Really wide fields of 10° can also capture other clusters – M23, M18, M28, M17 (the Horseshoe Nebula) along with a huge swath of the Sagittarius Milky Way. Longer exposures are required to bring out the beautiful detail of this area of the sky.

Located about 8° north of the spout of Sagittarius' teapot, M21 is found less than a degree northeast of M20 (The Trifid Nebula) and nearby M8 (the Lagoon Nebula) plus open clusters M23 and M28.

Suggested Instruments

3″+ refractor
6″+ reflector
3″+ catadioptric

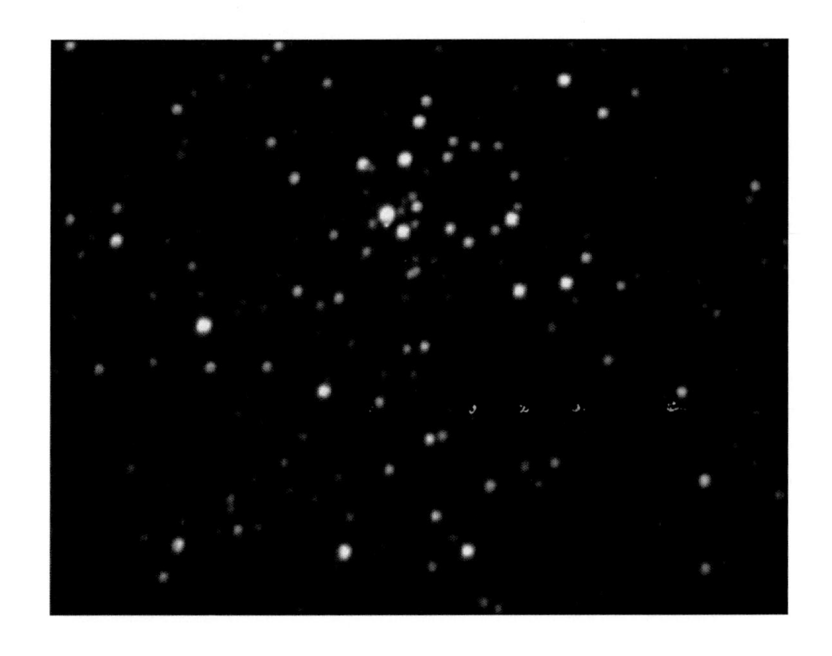

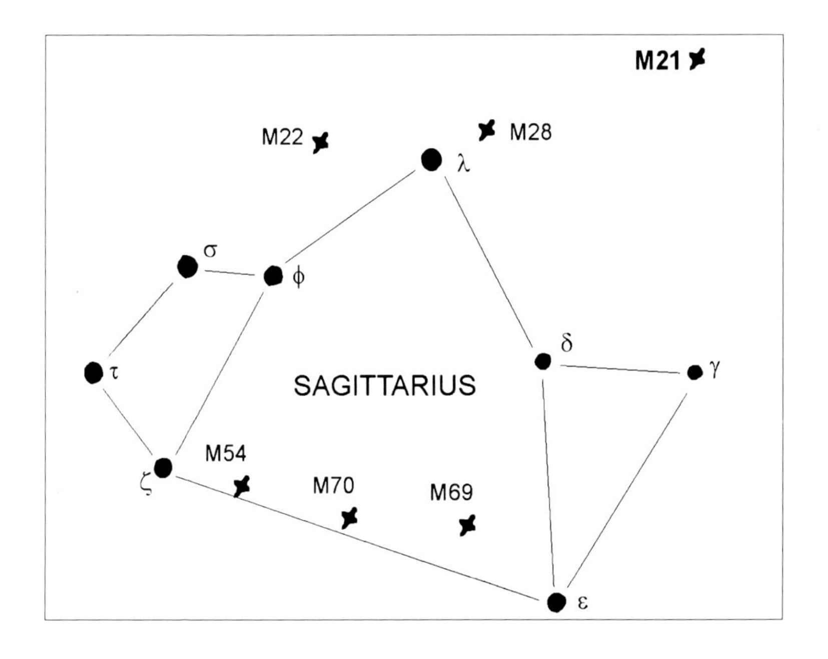
M21
M22
M28
λ
σ
φ
δ
τ
γ
SAGITTARIUS
M54
ζ
M70
M69
ε

NGC 6530 in Sagittarius

Type: Open cluster
Designation(s): NGC 6530
RA: 18h 04m
Dec: -24°20′
Visual magnitude: 4.5
Distance: 5,000 LY
Size: 35′

NGC 6530 is a beautiful cluster located and formed in the depths of the Lagoon Nebula (M8). It is very young and full bright, hot stars. In smaller telescopes it is quite beautiful, as it is nestled inside the associated nebulosity of M8. With a 4–5″ telescope, expect a nice view at about 60×. You can even bring the power up to 120× or so and look inside the cluster to see some of the interesting features deep within.

NGC 6530 was discovered by Hodierna in 1654. It is always been somewhat overlooked, as it is buried within one of the most spectacular nebulae in the heavens. However, it is important to note that it is really a part of M8, and as more stars form within M8 the radiation from the hot new stars will eventually disperse the nebula, leaving only the beautiful rich cluster behind.

The stars within NGC 6530 are only loosely attached gravitationally. This means that the cluster will disperse relatively quickly and will probably be almost nonexistent in less than a billion years. In the meantime, though, we can certainly enjoy it while it's here.

This cluster is perhaps 20 million years old and is still in the process of creating stars. The youngest stars seem to be around 100,000 years old. NGC 6530 is perhaps 30 light years across and growing as more stars are added. Of the 100+ members, at least a dozen have been identified as being variable. Many are of the Herbig Be or Ae type. These are stars that are still accumulating. There are unique emission lines present, as the stars shine through the disks of dust and gas beginning to form around them.

Photographically, NGC 6530 is extraordinary. The beautiful nebulosity surrounding it is stunning even with shorter exposures. There are numerous different aspects that can be captured, depending upon the framing and exposure. This is truly one of the more photogenic clusters in the heavens.

NGC 6530 is located in the midst of M8, a naked eye diffuse nebula, 6° north of γ (gamma) Sagittarii.

Suggested Instruments

3″+ refractor
6″+ reflector
3″+ catadioptric

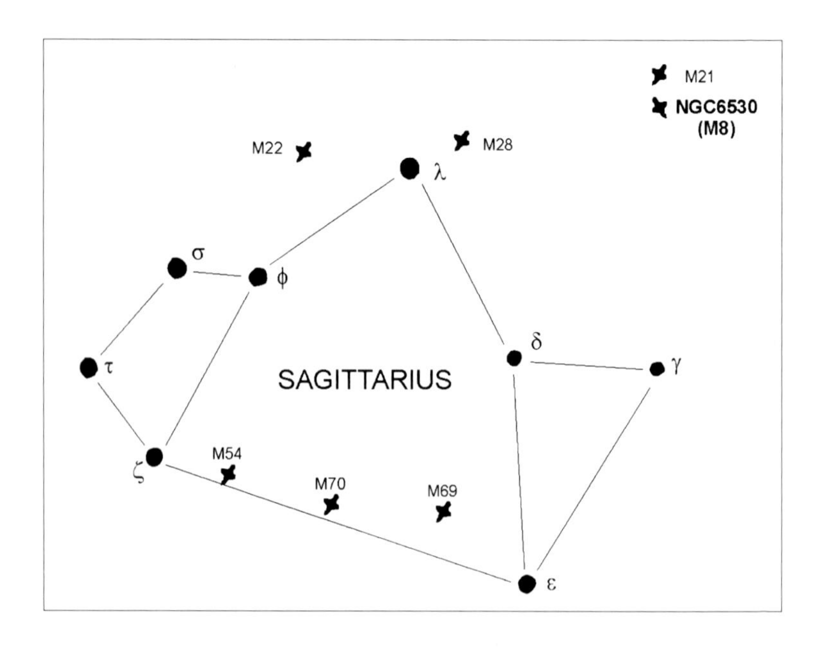
M21
NGC6530
(M8)
M22
M28
λ
σ
ϕ
δ
τ
γ
SAGITTARIUS
ζ
M54
M70
M69
ε

M24 in Sagittarius (Sagittarius Star Cloud)

Type: Open cluster
Designation(s): Messier 24, NGC 6603
RA: 18h 17m
Dec: -18°30′
Visual magnitude: 4.5
Distance: 16,000 LY
Size: 180′ (1°30′)

M24 is a magnificent cluster of stars containing thousands of stars. It is actually a piece of the spiral arm deeper in the galaxy. The cluster NGC 6603 is part of M24 and is particularly rich cluster located in the northern portion. This is a fine object in wide field telescopes or binoculars. With larger 'scopes and higher magnifications you can explore the myriad collections of stars among the structure within the star cloud.

M24 is huge cloud of stars about 600 light years across located about 16,000 light years distant in the direction of the center of the galaxy. There are several dark nebulae located in this object. This object is actually believed to be a "hole" in the obscuring matter between us and the center

of the galaxy. Without this obscuring matter the center of the galaxy would be perhaps 10 times brighter.

Photographically, this region is spectacular. Wonderful pictures can be made using piggyback astrophotography. Very nice shots are possible with lenses from 100 to 300 mm.

M24 is located just south of M17 and M18 in the northern portion of Sagittarius. There are many fine objects nearby, including clusters M23, M21, and M25 plus nebulae M8, M20, and M17.

Suggested Instruments

binoculars, finderscope
3″+ wide field refractor
wide field reflector

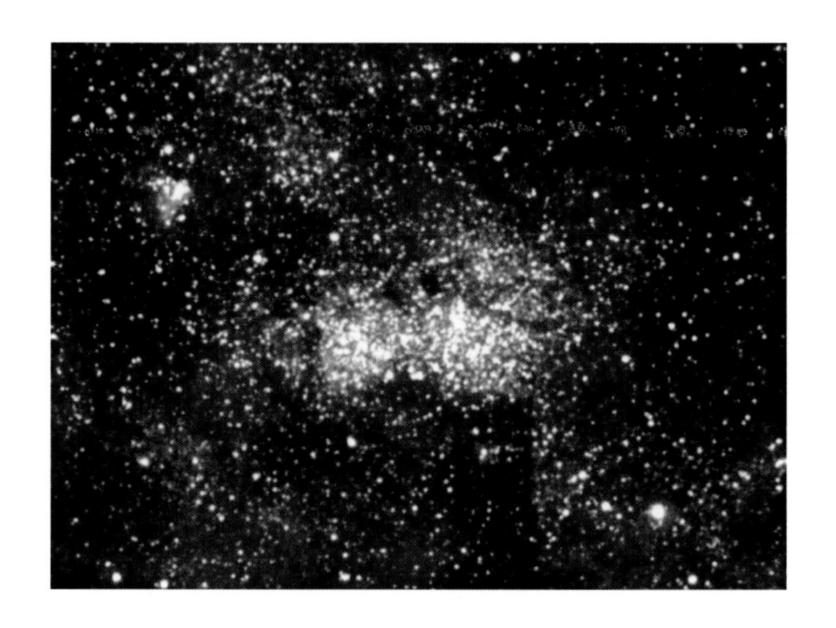

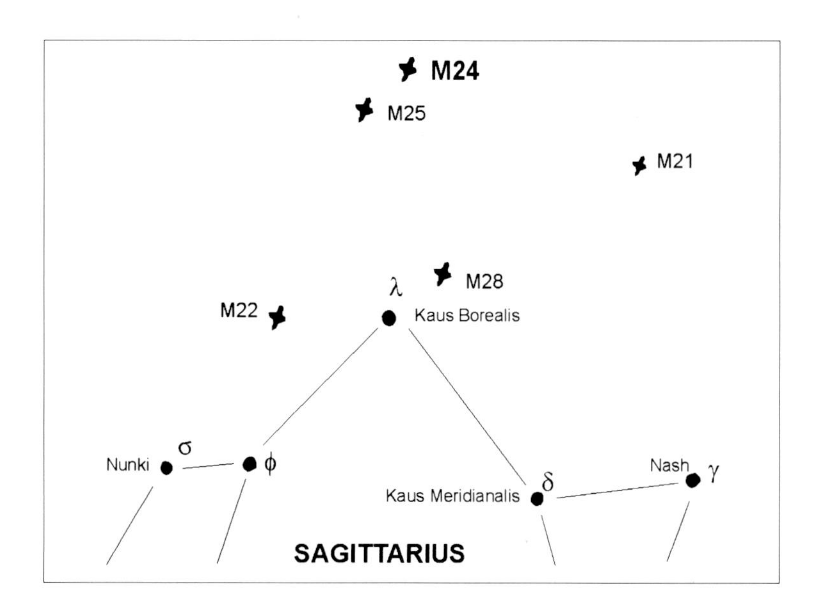
M24
M25
M21
M28
λ
M22
Kaus Borealis
σ
Nunki
ϕ
Nash
γ
δ
Kaus Meridianalis
SAGITTARIUS

M16 in Serpens (Eagle Nebula Cluster)

Type: Open cluster
Designation(s): Messier 16, NGC 6611
RA: 18h 19m
Dec: -13°46′
Visual magnitude: 6.4
Distance: 7,000 LY
Size: 25′

M16 is a vast swath of bright stars immersed in one of the most striking nebulae in the heavens. You need a telescope of at least 8″ to be able to detect the soft glow of the nebulosity. In larger 'scopes you can see the dark globules and filaments silhouetted on the nebulosity. These dark areas are contracting gas and dust, soon to become stars. M16 is truly a cluster in progress. The nebula has a relatively low surface brightness, so while it is faintly visible in a 6″ or 8″ telescope, apertures larger than 16″ are really wonderful for the viewing of its subtle beauty. Observers with larger 'scopes can view the grand pillars of dark nebulosity piercing the soft glow of the bright surrounding nebula. Dozens of brilliant blue stars radiate throughout to make a majestic impression on even the most seasoned observer.

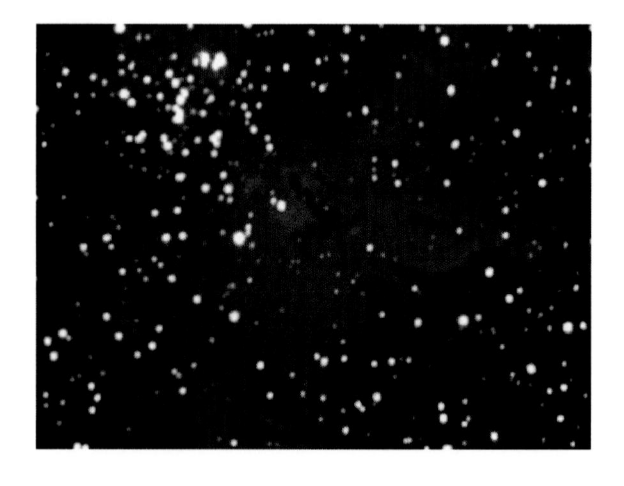

There are at least 80 stars in M16, with many more coming. The cluster and associated nebula is about 50 light years across and is located about 7,000 light years distant in the next spiral arm in from ours. This is a relatively young cluster at only 5 million years old. There are a large number of young, hot O (hot blue) stars in M16. These hot stars are exciting the gas in the nebula, causing it to glow. The brightest star has an absolute magnitude of -6.3, which makes it over 20,000 times more luminous than our Sun, a bright star indeed.

The brightest stars in the cluster shine with the light of tens of thousands of Suns. Such bright stars are very young and massive. These super-massive stars burn their hydrogen so fast, they will go novae and fade away to the obscurity of white dwarves in a relatively short time.

Early astronomers, including Charles Messier, observed the cluster to be immersed in the nebula. It is believed that de Cheseaux was first to discover it but it was rediscovered by Charles Messier later, who also discovered the nebula. For some reason William Herschel only recorded the star cluster. Since the NGC was based on his original work, it only records NGC 6611 as a star cluster and not as a nebula.

Photographic study of this cluster is possible with a wide variety of equipment and resulting detail. With shorter exposures and smaller telescopes, the cluster shines nicely in a soft glow of the nebula. With larger 'scopes and longer exposures, the details of nebula begin to come out along with many more stars. The size of the cluster and nebula are nearly ½ a degree, making the best photographs of the overall structure ones of a degree or so.

M16 is located in Serpens about 10° north of the top of Sagittarius' teapot. It is a few degrees north of the string of fine objects M24, M18, and M17 in Sagittarius. It is also located about 10° southeast of M11, a beautiful cluster in Scutum.

Suggested Instruments

3″+ refractor
6″+ reflector
3″+ catadioptric

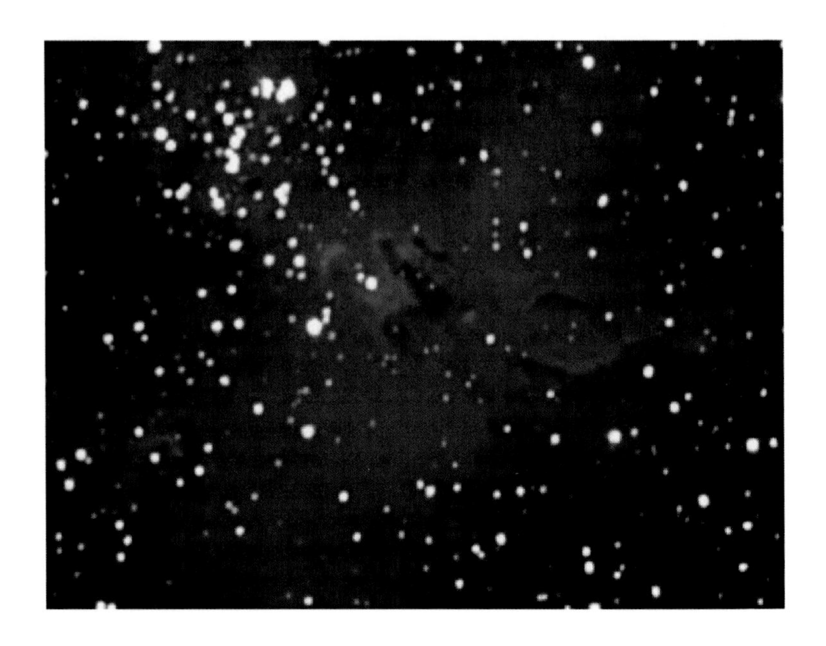

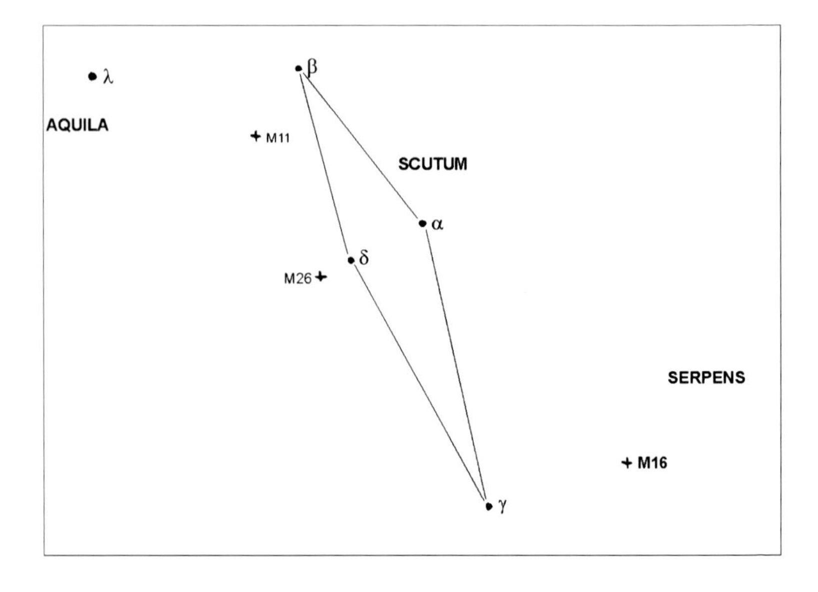
λ
AQUILA
β
M11
SCUTUM
α
δ
M26
SERPENS
M16
γ

M18 in Sagittarius

Type: Open cluster
Designation(s): Messier 18, NGC 6613
RA: 18h 20m
Dec: -17°08′
Visual magnitude: 7.5
Distance: 5,000 LY
Size: 10′

M18 is a small cluster of stars located on the northern border of Sagittarius. It is one of the less spectacular clusters in the heavens. However it does have a certain appeal. M18 is quite nice in a 3″ telescope, with more of its approximately 12 stars visible in larger 'scopes. It has a faint spiral shape.

M18 is a fairly young cluster of about 15 stars visible in most amateur telescopes. It probably contains over 50 stars in total. It is located roughly 5,000 light years toward the center of the galaxy. M18 contains bright yellow and bright blue stars. The Observatory of Geneva's color/magnitude diagram shows about 18 major members, most of which are bright blue stars. This indicates that this is a young cluster, about 30 million years old. There are also many yellow stars. There is still some of the cluster's

formation nebula present; however this is only visible in photographs made with large telescopes.

Upon discovering M18 in 1764 Messier described it as, "A cluster of small stars, a little below above nebula, No. 17, surrounded by slight nebulosity, this cluster is less obvious than the preceding, No. 16: with an ordinary telescope of 3.5-foot [FL], this cluster appears like a nebula; but with a good telescope one sees nothing but stars' (diam. 5′)."

There have been several photographic and spectroscopic studies done on M18. The results have varied somewhat, mainly measuring the brightest 20 stars or so.

M18 is located in the northern tip of Sagittarius near the border of Serpens and Scutum. It is just south of M17 (the Swan or Horseshoe Nebula), a probable star cluster birth place. M18 is also just north of M24, a huge star cloud comprised of many groups of stars. Other interesting objects nearby are the beautiful cluster M16 in Serpens and M25, a nice open cluster in Sagittarius.

Suggested Instruments

4″+ refractor
6″+ reflector
4″+ catadioptric

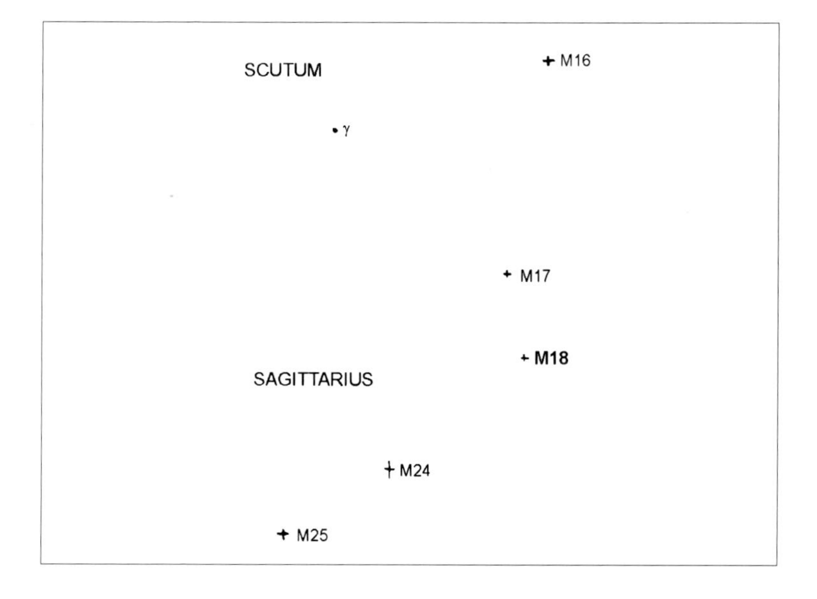
SCUTUM
M16
γ
M17
M18
SAGITTARIUS
M24
M25

M28 in Sagittarius

Type: Globular cluster
Designation(s): Messier 28, NGC 6626
RA: 18h 24m
Dec: -24°52′
Visual magnitude: 6.8
Distance: 18,000 LY
Size: 12′

M28 is a small but particularly rich globular cluster. Ideal for telescopes of at least 5", it is still certainly visible in smaller apertures. The cluster is somewhat overshadowed by nearby M8 and M22.

Messier discovered the cluster in July of 1764 and described it as "... seen with difficulty." John Herschel described it as a "fine object," showing the obvious differences between the capabilities of their instruments.

M28 is interesting in that it was one of the globular clusters studied closely by famous astronomer Helen Sawyer Hogg of the David Dunlop Observatory

in Ontario, Canada. She located many variable stars within this cluster, most of which were the RR Lyrae-type. RR Lyrae variables are stars that are commonly found in globular clusters. They pulsate very regularly and are used to accurately determine distances within the galaxy. Some of the variables in M28, interestingly enough, are irregular red variables, probably large dying stars going through the final contortions before they explode and fade into oblivion as white dwarfs.

M28 is located near the galactic core at about 18,000 light years distant and is more than 60 light years in diameter. It is moving very slowly relative to us, only about 1 km/s in recession.

Photographically, M28 is similar to many of the smaller globular clusters. It is easy to record on the photographic plate; however, for rich detailed views resolving a good number of stars, very painstaking photographic techniques are required.

Located just above the bright star λ (lambda) Sagittarii, it is very easy to find as λ (lambda) Sagittarii is the top star in the top of the Sagittarius teapot.

Suggested Instruments

4″+ refractor
6″+ reflector
5″+ catadioptric

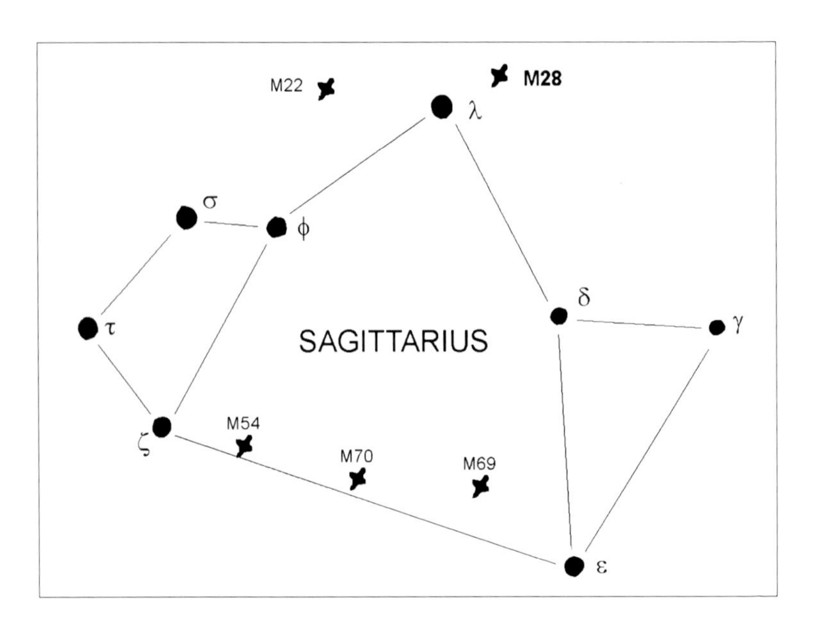
M22
M28
λ
σ
φ
δ
τ
SAGITTARIUS
γ
ζ
M54
M70
M69
ε

NGC 6633 in Ophiuchus

Type: Open cluster
Designation(s): NGC 6633
RA: 18h 28m
Dec: +06°34′
Visual magnitude: 4.7
Distance: 1,000 LY
Size: 28′

A large, bright cluster, NGC 6633 is a perfect target for telescopes of all sizes. Even with its few dozen stars, it shows amazing and interesting details. You can almost see swirls and lines of stars with patterns that change every time you look at it. You can view this particular cluster for hours and still find some new detail you never noticed before.

This large cluster is as large as the full Moon, and although it has relatively few stars, it is an interesting target. NGC 6633 is only around 600 million years old and continues to add large giant stars as its population ages. It is particularly nice in binoculars.

At only 1,000 light years distance, it is quite close. The distance and size indicates a diameter of around 8 light years.

There have been some recent studies done on the white dwarfs inside the cluster. So far seven have been discovered. One of these white dwarfs has been discovered to have the highest abundance of lithium of any such star yet found. The ramifications of this finding are still being discussed.

Wide field images of NGC 6633 can include the bright star Rasalhague for an interesting panorama, or close ups, as in the pictures in this book showing the loose knots of stars making it up.

NGC 6633 is quite easy to find, being just 14° east–southeast of Rasalhague. It is also near the interesting open cluster IC 4756.

Suggested Instruments

3″+ refractor
4″+ reflector
4″+ catadioptric

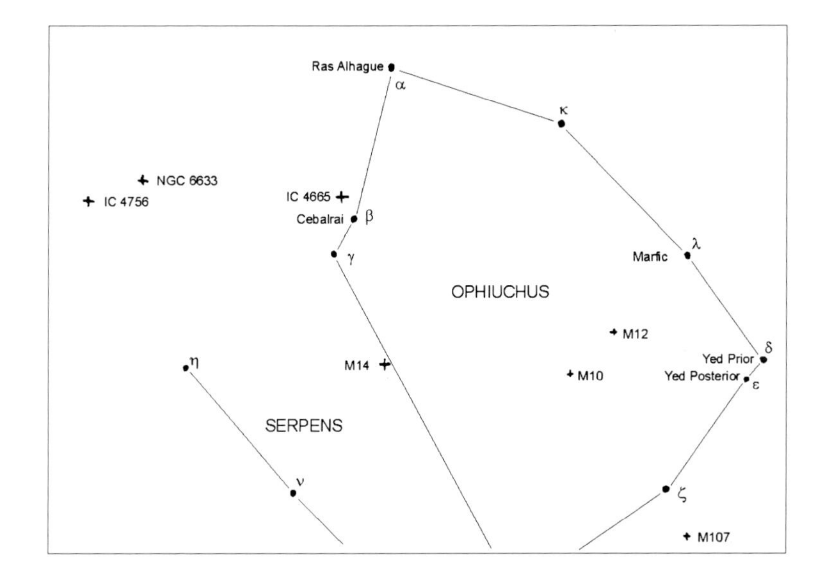
Ras Alhague
α
κ
NGC 6633
IC 4756
IC 4665
Cebalrai
β
γ
λ
Marfic
OPHIUCHUS
M12
δ
η
M14
Yed Prior
M10
Yed Posterior
ε
SERPENS
ν
ζ
M107

M25 in Sagittarius

Type: Open cluster
Designation(s): Messier 25, IC 4725
RA: 18h 32m
Dec: -19°16′
Visual magnitude: 6.0
Distance: 2,000 LY
Size: 40′

M25 is a good star cluster for binoculars or low power telescopes, as it is quite bright and more than ½ a degree in size. With a 3–4″ telescope you can see perhaps 30 or so of the brighter members. With a larger aperture of 8–10″ and 50× or 60× you can see the whole cluster and perhaps its 80 or so members. This is one of the nicer clusters in the heavens and helps to make Sagittarius one of the most cluster-rich regions of the sky.

M25 is believed to be at least 80 or 90 million years old, its distance of 2,000 light years and 40′ diameter giving it an actual size of nearly 25 light years. The central density is nearly 20 stars per cubic parsec, making it a rich, stable cluster. The lack of an NGC number is due to the fact that Herschel never included it in his *General Catalogue*, and subsequently it was omitted from the NGC. Messier did however include it in his catalog, and it is

believed to have been discovered as a cluster by de Cheseaux in 1745. It was added to the index catalog supplement to the NGC finally in the early twentieth century.

This cluster has many bright G-type stars and the noted Cepheid variable star U Sagittarii. Cepheid variables are not common in open star clusters. They are super-massive giant stars that pulsate very regularly to create changes in brightness and spectral class. U SGR changes in brightness from magnitude 6.2 to 7.1 over a period of 6.75 days. Cepheid variables are named for Delta Cephei, which was the first such star to be discovered. These stars help astronomers determine distances to remote galaxies, as they are extremely bright and there is a relationship between their luminosity and their period. Cepheid variables are usually spectral type A and F (white) stars. There is also a special type of short period Cepheid called cluster variables that are quite common in globular clusters. For more information on U Sagittarii and other variable stars, consult the AAVSO.

M25 is an interesting object photographically. Its large size makes it suitable for piggyback and prime focus photography. A 1,000–2,000 mm FL is ideal for this beautiful object.

M25 is located in Sagittarius about 5° north of the top of the teapot. It is located near clusters M18, M24, M21, M22, and M28, plus nebulae M8, M20, and M17. In fact with a 100 mm lens and a piggyback exposure, all these objects can be included in one shot.

Suggested Instruments

25×+ binoculars
2.4″+ refractor
4″+ reflector
3″+ catadioptric

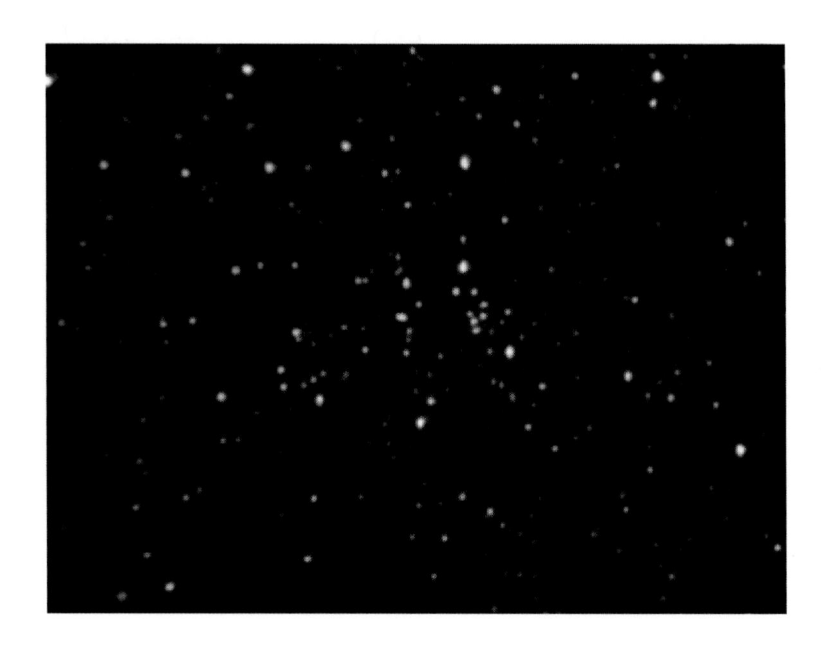

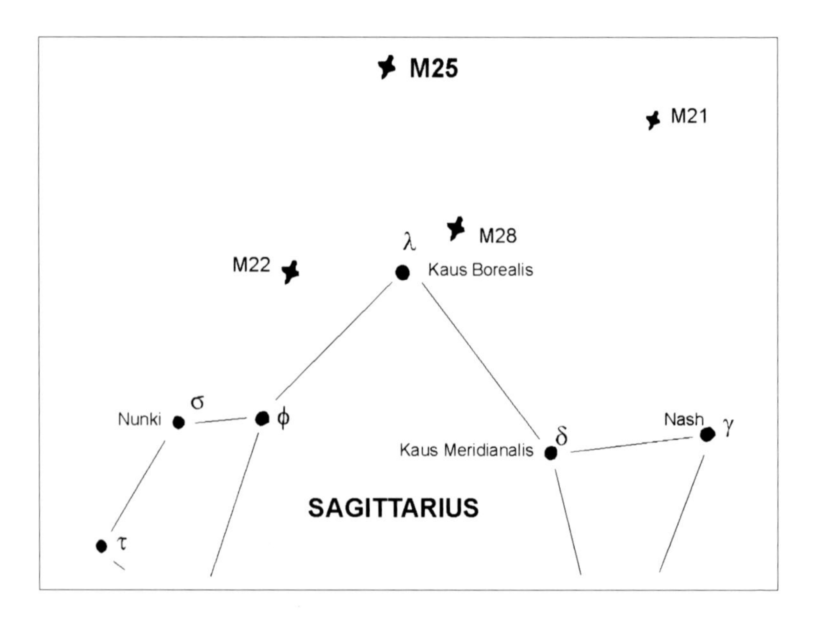
M25
M21
M28
λ
M22
Kaus Borealis
σ
Nunki
ϕ
Nash
γ
δ
Kaus Meridianalis
SAGITTARIUS
τ

M22 in Sagittarius

Type: Globular cluster
Designation(s): Messier 22, NGC 6656
RA: 18h 36m
Dec: -23°54′
Visual magnitude: 5.2
Distance: 10,000 LY
Size: 33′

M22 is one of the finest of all the globular clusters in the sky. It is the third brightest and definitely the brightest visible in the northern latitudes. It is certainly one of the most studied of all globular clusters.

Even though it is located nearly 10,000 light years away it is one of the brightest deep sky objects because it contains so many stars. More than 75,000 stars have been individually counted in M22, and certainly many more than that are uncountable. With a diameter of a bit over 50 light years, the core of this compact cluster has many stars packed per cubic light year.

There is a tiny planetary nebula at magnitude 14 inside M22, which has been determined to be a part of the cluster. It is a challenging object for an 18″ telescope and can even be elusive in larger 'scopes, depending upon

sky objects. The use of nebula filters and patience is helpful when attempting to view this object.

There have been some recent studies of events occurring in this cluster. These events have been characterized as possible large free floating planets, cosmic ray events, or perhaps nova bursts or flares. Continued study may reveal very interesting results about these wonderful objects.

As with many of the globular clusters there are a number of the RR Lyrae (or cluster)-type variable stars located within it.

There is no doubt that M22 is an extremely attractive target for imaging. The core is bluish with a dusting of red giants.

Suggested Instruments

25×+ binoculars
2.4″+ refractor
4″+ reflector
3″+ catadioptric

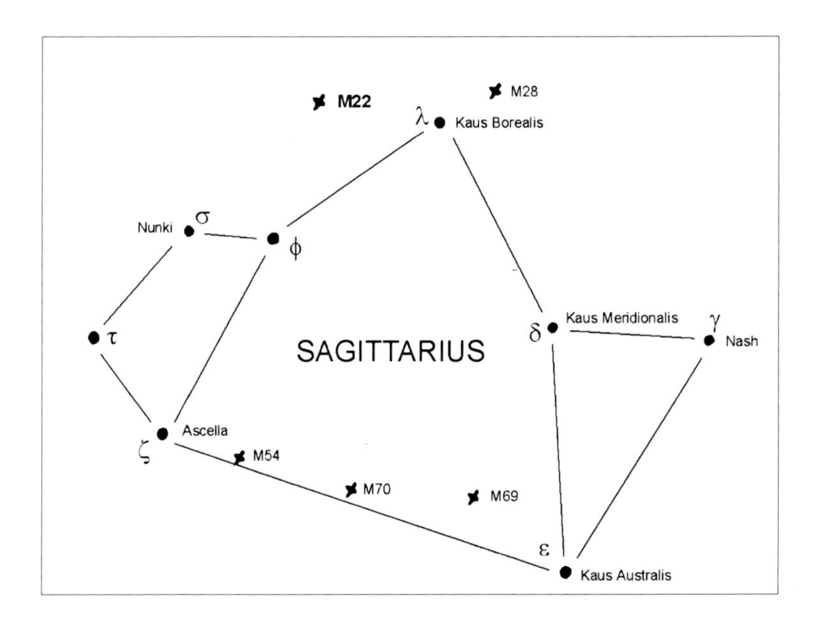
M22
M28
Kaus Borealis
λ
Nunki
σ
φ
τ
SAGITTARIUS
Kaus Meridionalis
δ
γ
Nash
Ascella
ζ
M54
M70
M69
ε
Kaus Australis

M26 in Scutum

Type: Open cluster
Designation(s): Messier 26, NGC 6694
RA: 18h 45m
Dec: -9°23′
Visual magnitude: 8.0
Distance: 5,000 LY
Size: 15′

M26 is a tight cluster, which is rather dim and needs a larger telescope to appreciate its beauty. It is one of the nicer clusters in structure and density. With a refractor of 5″ or more or an 8″ or bigger reflector this cluster really shows its true beauty. Refractor views of this cluster are especially noteworthy, as the added contrast really brings out the central groupings really well.

Most of the stars in this 90-million-year-old cluster seem to be spectral class B (blue) stars. With a 5″ or 6″ refractor or an 8″ reflector, about 25 or 30 of the cluster's almost 100 members become visible. In larger 'scopes many more of the fainter stars resolve. M26 is located about a light year's distant and is about 15 light years in diameter.

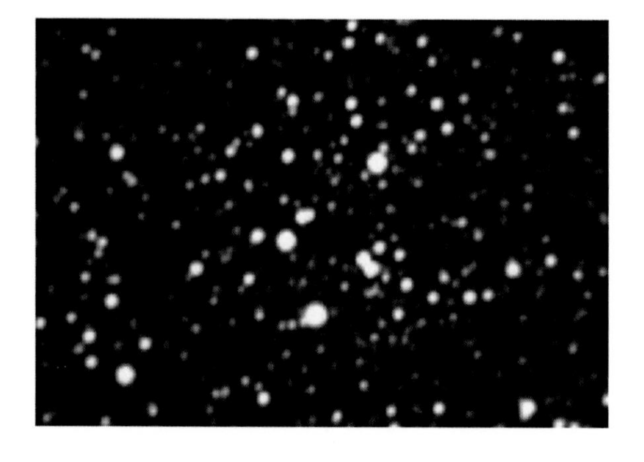

M26 is located less than a degree away from the famous variable star δ (delta) Scuti. Stars of this type are related to the Cepheid variables (see notes on U Sagittarii in M25). This star is pulsating at the extremely consistent rate of once every 279 min. The actually expands and contracts in a regular rhythm, varying in brightness by approximately 1/6 magnitude. This star is of spectral type F (white), is about 30 times the luminosity of the Sun. It is located about 250 light years from us. More than 250 of this class (called Delta Scuti variables) have been identified. For more information on Delta Scuti and other variable stars, consult the AAVSO.

M26 is a fine photographic target. Long exposures at longer effective focal lengths (3,000 mm+) are best. The catadioptric telescopes of 8″+ provide very nice results, as do the longer focal length refractors and reflectors.

M26 is located about 4° south of bright open cluster M11 and about 1° west of δ (delta) Scuti.

Suggested Instruments

4″+ refractor
6″+ reflector
8″+ catadioptric

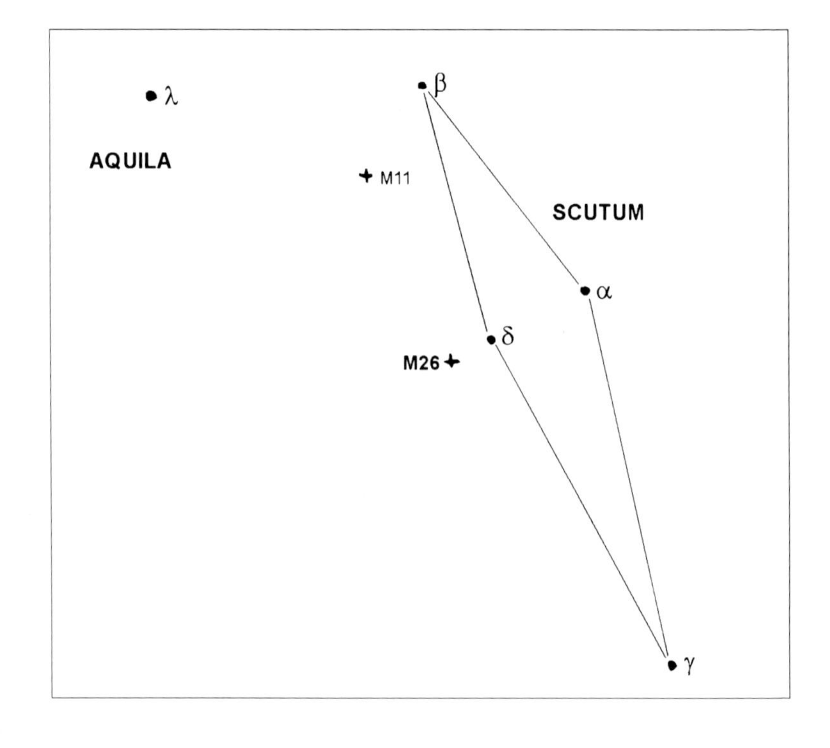
λ
AQUILA
β
M11
SCUTUM
α
δ
M26
γ

M11 in Scutum (Wild Duck Cluster)

Type: Open cluster
Designation(s): Messier 11, NGC 6705
RA: 18h 51m
Dec: -06°16′
Visual magnitude: 6.3
Distance: 5,500 LY
Size: 20′

M11 is a particularly rich star cluster located in the Scutum star cloud. It is a beautiful patch of stars at 150× in a 4–8″ telescope. In larger 'scopes more and more stars resolve deeper in the center of the cluster.

M11 is estimated to lie about 5,500 light years distant and somewhat nearer to the galactic center than us. It contains close to 3,000 stars, with a density in the central third of nearly 100 stars per cubic parsec. M11 has a diameter of about 20 light years. Many of the stars are young giant stars around 100 times as luminous as the Sun. If the Sun were in M11 we would see hundreds of first magnitude stars with quite a few many times brighter than Sirius. The best estimates for M11's age put it at around 500 million

years. There are many very luminous stars of the B and A (blue) class, with several dozen red and yellow giants. The contrast is nicely visible in larger reflecting telescopes, while the contrast of a 6″+ refractor really brings out the central stars beautifully.

The cluster is located near R Scuti, an RV Tauri-type variable star that ranges from a magnitude of 4.5 to around 9 in a complex irregular period of around 146 days. The seventh magnitude double star Σ 2391 is located between R Scuti and M11.

Nice photographs of this cluster can be done with 6″+ telescope which, depending on exposure, will reveal various personalities. With moderate-exposure CCD imaging, this cluster reveals some incredible details.

M11 is can be found in Scutum about 20° north of Sagittarius and about 2° southeast of β Scuti. About 5° south is the smaller open cluster M26.

Suggested Instruments

3″+ refractor
6″+ reflector
3″+ catadioptric

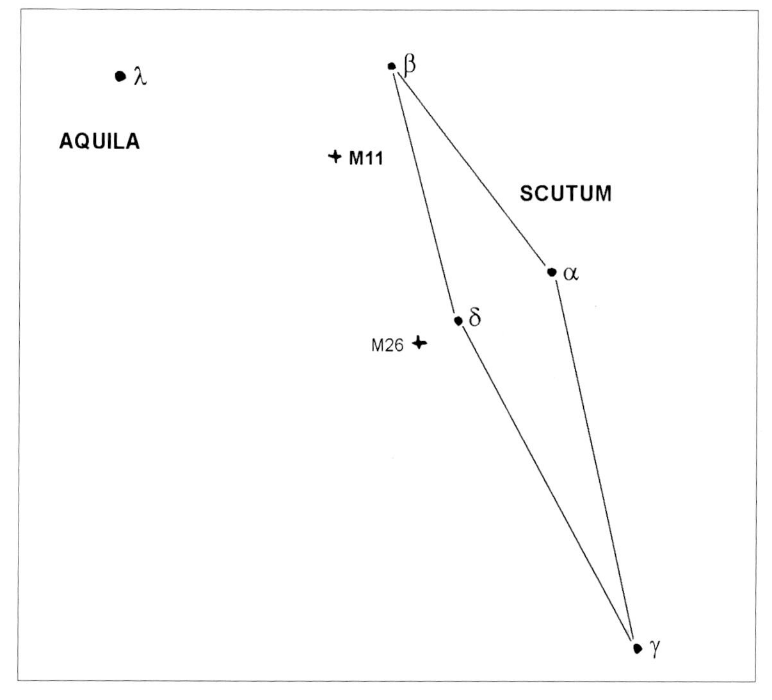
λ
AQUILA
β
M11
SCUTUM
α
δ
M26
γ

M56 in Lyra

Type: Globular cluster
Designation(s): Messier 56, NGC 6779
RA: 19h 16m
Dec: +30°11′
Visual magnitude: 8.4
Distance: 30,000 LY
Size: 7′

M56 is a somewhat elliptical-shaped globular cluster. Even though it is quite distant, 30,000 light years, it is fairly bright due to its tightly concentrated core. The central portion of this cluster is more than 85 light years across and probably contains more than 100,000 stars.

M56 contains relatively few variables. These few variables are accessible for study with relatively small telescopes of perhaps 8–12″ or more.

M56 was actually discovered by Messier himself in 1779. While he was unable to resolve it in his primitive telescope, he nonetheless determined it was a non-cometary object from its lack of motion.

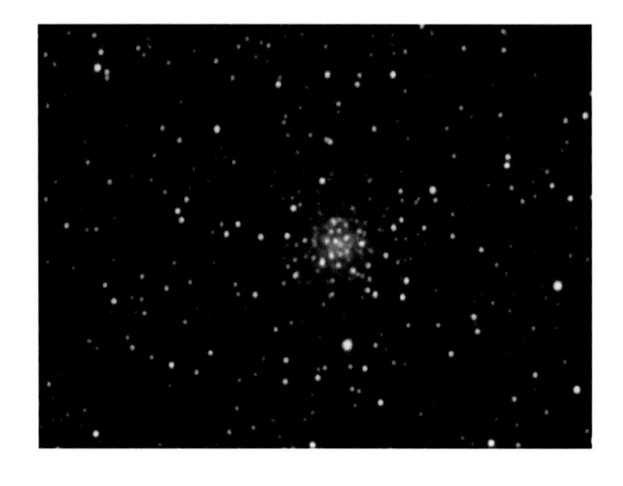

Imaging M56 is a lot of fun, as it contains a hazy bluish center sprinkled with yellow and orange stars.

Suggested Instruments

4″+ refractor
6″+ reflector
6″+ catadioptric

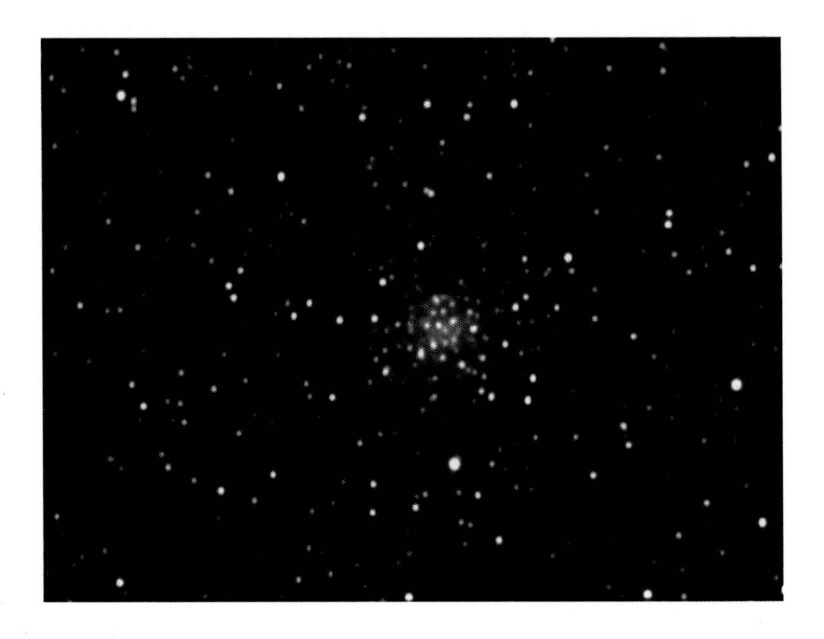

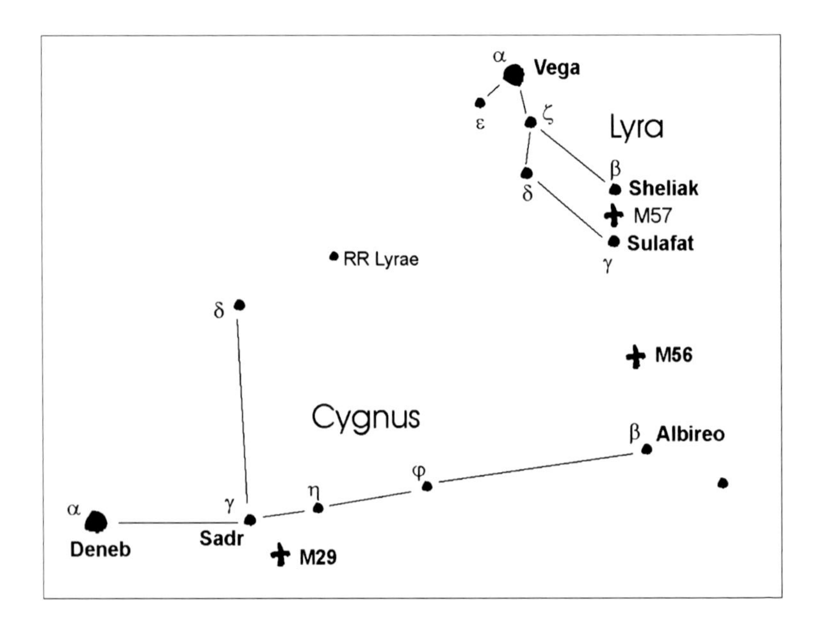
α
Vega
ε
ζ
Lyra
β
Sheliak
δ
M57
Sulafat
γ
RR Lyrae
δ
M56
Cygnus
β
Albireo
φ
η
γ
α
Deneb
Sadr
M29

Index

A

Algenib, 33, 34
Almach, 25
Andromeda, 24–26
Angular momentum, 14
Antares, 91, 94, 97, 115
Aperture, 15, 17, 51, 84, 96, 100, 105, 108, 120, 156
Arcturus, 82, 85
Asymmetry, 112
Atlas Coeli, 11
Auriga, 45–53
Aztec, 37

B

Beehive, 27, 67, 69–71
Be-Shell Stars, 37
Big Dipper, 75, 76
Black hole, 9, 14, 15, 112
BU Puppis, 64
Butterfly Cluster, 126–128

C

Cancer, 67, 69–74
Canes Venatici, 84–86
Canis Major, 42, 60–62
Cassiopeia, 21–23, 27
CCD, 15, 25, 28, 34, 37, 67, 71, 73, 79, 88, 91, 94, 97, 100, 109, 115, 118, 124, 133, 166
Celabra, 124
Cepheid, 13, 82, 94, 106, 115, 157, 163
Chandrasekhar, S., 14
Cluster, 3, 9–13, 17, 21–170
Collinder, 75–77
Collinder 285, 75–77
Coma Berenices, 81–83
Cor Caroli, 85
Corvus, 79

D

De Cheseaux, P.L., 93, 145, 157
Declination, 105
De Lacaille, N.L., 11
Density, 10, 14, 31, 90, 100, 132, 136, 156, 165
Doppler, 7
Double Cluster, 27–29
Dreyer, J.L.E., 3, 112

E

Eagle Nebula, 144–146
Eichner, Laurits Dan, IX
Einstein, Albert, IX
Electron7, 13, 14
Elmer, Charles W., IX

F

Fitz, Harry, IX

G

Galaxy, 3, 5, 9–12, 25, 42, 43, 55, 67, 79, 85, 88, 100, 103, 118, 136, 141, 142, 147, 151

C.A. Cardona III, *Star Clusters: A Pocket Field Guide,* Astronomer's Pocket Field Guide, DOI 10.1007/978-1-4419-7040-4,

Gamma ray, 15, 112
Gemini, 54–56
Giza, 45
Globular cluster, 3, 5, 9, 12, 13, 15, 42–44, 73, 78–104, 108–125, 130, 133, 150–152, 157, 159–161, 168–170
Graffias, 91
Gravitational, 3, 5, 12, 25, 27, 139

H
Halley, E., 11, 99
H-alpha, 58
Harvard, 6, 85, 88
Helium, 5, 6, 9, 13
Herbig, 139
Herbig-Haro, 58
Hercules, 99–101, 103, 117–119
Herschel, J., 99, 103, 124, 150, 156
Herschel, W., 11, 12, 24, 66, 72, 145
Hertzsprung, E., 7
Hess, Fred, IX
Hevelius, J., 11, 87
Hipparcos, 27
Hodierna, G.B., 24, 33, 106, 138
Hogg, H.S., 150
Horseshoe Nebula, 133, 136, 148
Hubble, 91
Hyacinthus, 39
Hyades, 3, 10, 39–41, 70
Hydra, 66–68, 78–80

K
Kimah, 37
Kirch, G., 87
Koehler, J.G., 72

L
Lacaille, A., 11, 12
Lagoon Nebula, 133, 135, 136, 138
Lepus, 42–44
Lippincott, Sara Lee, IX
Lithium, 154
Luminosity, 10, 13, 28, 46, 48, 60, 67, 70, 81, 82, 85, 94, 112, 124, 163
Lyra, 168–170

M
M4, 91, 93–95
M5, 87–89
M7, 127, 129–131
M10, 103, 108–110, 121
M11, 145, 163, 165–167
M13, 78, 85, 99–101, 103, 109, 117, 118
M18, 133, 136, 142, 145, 147–149, 157
M19, 114–116
M21, 133, 135–137, 142, 157
M22, 136, 150, 157, 159–161
M23, 132–134, 136, 142
M24, 133, 141–143, 145, 148, 157
M25, 142, 148, 156–158, 163
M28, 136, 150–152, 157
M34, 30–32
M35, 54–56
M37, 51–53
M38, 45–48
M44, 27, 67, 69–71, 73
M48, 66–68
M53, 81–83
M56, 168–170
M62, 111–113, 115
M67, 70, 72–74
M68, 78–80
M79, 42–44
M80, 90–92, 94
M92, 117–119
M103, 21–23
M107, 96–98
Magellanic, 12
Main sequence, 8, 12, 25, 51, 54, 69, 72, 79, 126, 127, 132
Mechain, P., 42, 96
Melotte, 33–35, 39–41

Melotte 20, 33–35
Melotte 25, 39–41
Messier, C., 3, 10–12, 42, 63, 66, 72, 85, 88, 91, 93, 94, 99, 103, 106, 112, 115, 121, 124, 145, 148, 150, 156, 168
Metals, 9, 12, 63, 73
Mira, 79, 109
Mizar, 76
Monoceros, 57–59

N
Nebulae, 9–12, 57, 96, 120, 121, 138, 141, 142, 144, 157
Neutron star, 9, 14, 112
NGC 752, 24–26
NGC 2158, 55
NGC 2244, 57–59
NGC 6231, 105–107
NGC 6530, 138–140
NGC 6633, 153–155
Nova, 27, 34, 88, 91, 106, 124, 160
Novae, 9, 91, 145

O
OB association, 33, 34
Open cluster, 21, 24, 30, 33, 36, 39, 45, 48, 51, 54, 57, 60, 63, 66, 69, 72, 73, 75, 105, 126, 129, 130, 132, 133, 135, 136, 138, 141, 144, 147, 148, 153, 154, 156, 162, 163, 165, 166
Ophiuchus, 96–98, 102–104, 108–116, 120–125, 153–155

P
Palomar, 88
Paris Observatory, 42
Perseus, 27–35
Planetary nebulae, 9, 79, 159
Pleiades, 3, 10, 27, 36, 37, 48
Pleione, 37
Population I, 9
Population II, 12
Praesepe, 67, 69–71
Ptolemy, 69, 129–131
Pulsar, 13–15, 94, 112
Puppis, 63–65

Q
Quasars, 106

R
Rasalhague, 154
Ras al Muthallath, 25
Red giant, 9, 13, 22, 51, 52, 60, 61, 67
Right Ascension, 17
ROSAT, 112
RR Lyrae, 12, 13, 17, 79, 81, 85, 88, 94, 97, 100, 112, 115, 118, 124, 151, 160
Ruchbah, 22
Russell, H.N., 7

S
Sabik, 121
Sagittarius, 127, 132–143, 145, 147–152, 156–161, 166
Scorpius, 90–95, 105–107, 126–131
Scutum, 145, 148, 162–167
Segin, 22
Serpens, 87–89, 144–146, 148
Shapley, H., 11
Solar, 6, 28, 63, 70, 75
Spectra, 6, 7, 9
Spiral,5, 9, 10, 34, 103, 124, 141, 145, 147
Sumerians, 45

T
Taurus, 36–40, 70
Tianquiztli, 37
Trifid Nebula, 133, 135, 136
Trumpler, R.J., 10
T Tauri, 40

U
URSA Major, 75–77

V
Van De Kamp, Peter, IX
Variable, 12, 37, 40, 64, 70, 79, 85, 87, 88, 94, 97, 108, 109, 115, 118, 124, 127, 133, 139, 151, 157, 160, 163, 166

W
White dwarf, 9, 14, 40, 55, 69, 145, 151, 154,
Wild Duck Cluster, 165–167
Wisniewski, J., 61, 67, 71, 100

X
X-ray, 15, 37, 112, 127